HANNAH ARENDT
AND HUMAN RIGHTS

T0323840

Hannah Arendt & Human Rights

The Predicament of Common Responsibility

PEG BIRMINGHAM

Indiana University Press
Bloomington & Indianapolis

This book is a publication of

Indiana University Press
601 North Morton Street
Bloomington, IN 47404-3797 USA

http://iupress.indiana.edu

Telephone orders   800-842-6796
Fax orders   812-855-7931
Orders by e-mail   iuporder@indiana.edu

The paper used in this publication meets the minimum requirements of
American National Standard for Information Sciences—Permanence of
Paper for Printed Library Materials, ANSI Z39.48-1984.

MANUFACTURED IN THE UNITED STATES OF AMERICA

Library of Congress Cataloging-in-Publication Data

Birmingham, Peg, date
  Hannah Arendt and human rights : the predicament of
common responsibility / Peg Birmingham.
     p.   cm. — (Studies in Continental thought)
  Includes bibliographical references and index.
  ISBN 978-0-253-21865-0 (pbk. : alk. paper) 1. Arendt, Hannah.
2. Human rights—Philosophy. 3. Responsibility. I. Title. II. Series.
JC251.A74B57 2006
323.092—dc22
                                                    2006006223

2 3 4 5 6   12 11 10 09 08 07

For Clare

We become aware of the existence of a right to have rights (and that means to live in a framework where one is judged by one's actions and opinions) and a right to belong to some kind of organized community, only when millions of people emerge who had lost and could not regain these rights because of the new global political situation.

Hannah Arendt,
The Origins of Totalitarianism

CONTENTS

ACKNOWLEDGMENTS

I am profoundly grateful to Professor Fred Kersten, who first introduced me to the world of philosophy, most notably Husserlian phenomenology. His stories of studying with Arendt, Gurwitsch, and Jonas at the New School for Social Research first led me to these thinkers, especially Arendt. The years I studied with him at the University of Wisconsin in Green Bay constituted my archaic beginning that continues to animate the present.

I owe a great debt to my students at DePaul University, whose contributions to the seminars I have given on Arendt helped me better formulate the arguments in this book. It has been my good fortune to have David F. Krell as a colleague and friend. Not only did he read a penultimate draft of this manuscript, making copious suggestions and comments, but he has been enormously encouraging of my work since our first disagreement over Heidegger's thought nearly two decades ago. Of course, any errors and lapses of judgment are entirely my own. Conversations with colleagues Bill Martin, Darrell Moore, Will McNeill, Tina Chanter, Michael Naas, and Elizabeth Rottenberg have been immensely helpful. A word of thanks to Bernie Flynn, who graciously accepted my invitation to give a course on Hannah Arendt at the Collegium Phaenomenologicum in 1990 and with whom I have had many subsequent insightful conversations on Arendt's thought. Thanks also to Robin May Schott, who was in the audience when I presented a paper on Arendt and the banality of evil at the Society for Phenomenology and Existential Philosophy in 2001 and encouraged me to submit it for consideration in a special volume of *Hypatia* she was editing on feminist philosophy and the problem of evil. The framework of this book grew out of my contribution to that volume.

Ben and Laura Nicholson have provided the respite of friendship, their dinner table a place in which the pleasure of company flourishes. I am also grateful to Richard Tobin, who was there in the beginning and continues to be so. Dee Mortensen, Elisabeth Marsh, and Kate Babbitt at Indiana University Press have been wonderful editors. Kate Babbitt's meticulous copyediting has made this a more elegant book.

And finally, I would like to thank Dean Michael L. Mezey for his unfailing support and his insistence that the duties of chairing the philosophy department not take priority over the completion of this work. Research for this book was supported by a DePaul University Summer Research Grant as well as a two-quarter research leave, both extremely helpful in giving me the time needed to finish this work.

I dedicate this book, with love, to my daughter Clare.

Grateful acknowledgment is given to the following sources for permission to quote from previously published material:

Research in Phenomenology, Volume 33, 2003—Peg Birmingham, "The Pleasure of Your Company: Arendt, Kristeva, and an Ethics of Public Happiness," pp. 55–72.

Revolt, Affect, Collectivity: The Unstable Boundaries of Kristeva's Polis, ed. Tina Chanter and Ewa Plonowska Ziarek, pp. 129–41. Reprinted by permission of the State University of New York Press. © 2005 State University of New York. All Rights Reserved.

Hypatia, Volume 18, Number 1, Winter 2003, "Holes of Oblivion: The Banality of Evil," pp. 80–89.

ABBREVIATIONS USED
IN THE TEXT FOR BOOKS
BY HANNAH ARENDT

BPF	*Between Past and Future* (New York: Penguin, 1977)
EJ	*Eichmann in Jerusalem: A Report on the Banality of Evil* (New York: Penguin Books, 1963)
EU	*Essays in Understanding,* edited by Jerome Kohn (New York: Harcourt Brace and Company, 1994)
HC	*The Human Condition* (Chicago: University of Chicago Press, 1958)
JP	*The Jew as Pariah: Jewish Identity and Politics in the Modern Age,* edited by Ron H. Feldman (New York: Grove Press, 1978)
KPP	*Lectures on Kant's Political Philosophy,* edited and with an interpretive essay by Ronald Beiner (Chicago: University of Chicago Press, 1982)
LMT	*Life of the Mind,* vol. 1, *Thinking* (New York: Harcourt Brace Jovanovich, 1978)
LMW	*Life of the Mind,* vol. 2, *Willing* (New York: Harcourt Brace Jovanovich, 1978)
LSA	*Love and Saint Augustine,* edited by Joanne Vecchiarelli Scott and Judith Chelius Stark (1929; reprint, Chicago: University of Chicago Press, 1996)
MDT	*Men in Dark Times* (New York: Harcourt, Brace & World, 1968)
OR	*On Revolution* (New York: Penguin Books, 1963)
OT	*The Origins of Totalitarianism* (New York: HBJ, 1951)
OV	*On Violence* (New York: Harcourt, Brace & World, 1970)
PP	*The Promise of Politics,* edited by Jerome Kohn (New York: Schocken Books, 2005)
RJ	*Responsibility and Judgment,* edited by Jerome Kohn (New York: Schocken Books, 2003)
RV	*Rahel Varnhagen, The Life of a Jewish Woman,* translated by Richard and Clara Winston (New York: Harcourt Brace Jovanovich, 1974)

HANNAH ARENDT
AND HUMAN RIGHTS

Introduction: The Problem
of Human Rights

Hannah Arendt's most important contribution to political thought may be her well-known and often-cited notion of the right to have rights. She first articulated the idea in *The Origins of Totalitarianism* in the context of her analysis of the decline of the nation-state. Its eventual dénouement in the death camps, she argues, could have happened only because of a philosophically invalid and politically impotent notion of human rights. Arendt's entire work can be read as an attempt to work out theoretically this fundamental right to have rights. Yet her notion of the right to have rights remains the least understood aspect of her work. While many of her most astute and careful readers refer admiringly to the phrase, they criticize her for providing no theoretical justification for this right. For example, Benhabib argues that Arendt can "ultimately offer no philosophical justification either for her belief in universal human rights or for the category of crimes against humanity."[1] Benhabib asks:

> Is the whole category of "human rights," the "existence of a right to have rights," in her perspicacious phrase, a defensible one? Do human beings have rights in the same way in which they can be said to have body parts? If we insist that we must treat all humans as being entitled to the right to have rights, on the basis of which philosophical assumptions do we defend this insistence? Do we ground such respect for universal human rights in nature, in history, or in human rationality? One searches in vain for answers to these questions in Arendt's text. But by withholding a philosophical engagement with the justification of human rights, by leaving ungrounded her own ingenious formulation of the "right to have rights", Arendt also leaves us with a disquiet about the normative foundations of her own political philosophy.[2]

Dana Villa offers a similar critique, arguing that Arendt "devotes little attention to the liberal tradition and the theory of rights which animates it."[3] He claims that Arendt eschews a theory of human rights in favor of political action which, in turn, renders her thought relatively "unconcern[ed] with the topic of justice."[4]

In a similar vein, Margaret Canovan argues that "however committed she herself might be to the ideas of equal human worth and equal human rights, [Arendt] certainly did not suppose that this was something that could be demonstrated or deduced from human plurality."[5] Likewise, Claude Lefort argues that Arendt's claim that human rights are derived from "the fiction of human nature" and her argument that rights are nothing other than the rights of citizens make it impossible for her to provide a philosophical basis for human rights that would demand that the mutual recognition of individuals made in one another's likeness extend beyond the gates of the city. Because it lacks such a basis, Lefort argues, it is difficult to see in Arendt's thought "how we can possibly justify our condemnation of totalitarianism, except on the crude and almost accidental grounds that its conquests are a threat to our society."[6]

At the same time that Arendt is criticized for providing no theoretical formulation of human rights, however, thinkers such as John Rawls and Michael Ignatieff argue that such a project is either unnecessary (Rawls) or futile (Ignatieff), a foundational longing tied to universal Enlightenment ideals that cannot be theoretically substantiated in this post-metaphysical multicultural age. Rawls argues that it is unnecessary, even counterproductive, to articulate a foundation for human rights and that to try to do so is to reintroduce a notion of the good into a political space that is better off without it. The liberal tradition, he argues, provides its own *historical* foundation. And that is good enough.[7]

While Ignatieff refers approvingly to Arendt's notion of the right to have rights, he nonetheless argues that her notion, or any other notion of human rights, cannot be given a universal foundation other than the theological: "If the purpose of human rights is to restrain the human use of power, then the only authority capable of doing so must lie beyond humanity itself, in some religious source of authority."[8] Because they lack a theological basis, he argues, theoretical claims concerning human rights are always unclear and controversial; it is better to move to the prudential, forgoing the question of why we have rights in favor of recognizing what rights actually do for us. Embracing Isaiah Berlin's insight that rights are founded not on reason but on the memory of horror, Ignatieff argues that "all that can be said about human rights is that they are necessary to protect individuals from violence and abuse, and if . . . asked why, the only possible answer is history."[9] Indeed, Ignatieff argues that any theory that claims more than this falls into "rights idolatry" in which conditional faith in our species becomes an idolatrous worship of the purely human: "Why, exactly, do we think that ordinary human beings, in all their radical heterogeneity of race, creed, education, and attainment, can be viewed as possessing the same equal and inalienable rights? If idolatry consists in elevating any purely human principle into an unquestioned absolute, surely human rights looks like an idolatry."[10]

The task of the present volume is threefold. First is to argue that readers of Arendt have failed to grasp that one of her primary concerns, beginning with *The Origins of Totalitarianism,* is the working out of a theoretical foundation for a reformulation of the modern notion of human rights. Arendt's reformulation, I submit, is rooted in a principle of common humanity that does not fall into an idolatrous worship of the human. Self-described as one who has joined the ranks of post-metaphysical thinkers, Arendt has formulated a notion of a common humanity that is rooted not in an autonomous subject or in nature, history, or god; instead, she finds this principle in the anarchic and unpredicable event of natality. For Arendt, the event of natality, with its inherent principle of humanity, provides the ontological foundation for human rights. Moreover, the archaic event of natality carries two principles: the principle of *initium* and the principle of givenness. In chapters 2 and 3, respectively, I take up each aspect of this stratified principle of humanity. Finally, in chapter 4, I examine the affectivity at work in the event of natality and its stratified principle, arguing that this affectivity is more complicated and nuanced than even Arendt herself—in spite of her sober-minded view of human existence—was able to admit. Here I take up Julia Kristeva's recent discussion of Hannah Arendt's thinking, arguing that it illuminates Arendt's unwavering insistence on the predicament of common responsibility inherent in the event of natality itself. For Arendt, and this is the pessimism present throughout her work, the predicament of common responsibility is our capacity for both horror and gratitude, both violence and pleasure, when confronted with our common humanity in the company of others. Yet this predicament, which is inherent in the human condition, does not discourage Arendt from seeking a foundation for the right of human beings born to have rights.

In the conclusion I briefly address the difficult issue of the political institution of the rights to have rights. Arendt's writings on post–World War II Europe and her essays on Palestine in the 1940s and early 1950s provide insights for rearticulating notions of sovereignty and the nation-state, which, in turn, allow us to consider the kinds of immigration, naturalization, and citizenship practices that would be compatible with an Arendtian commitment to the right to have rights.

ONE

The Event of Natality:
The Ontological Foundation
of Human Rights

> Mankind, whether a religious or humanistic ideal, implies
> a common sharing of responsibility. The shrinking of
> geographic distances made this a political actuality of the
> first order. It also made idealistic talk about mankind and
> the dignity of man an affair of the past simply because all
> these fine and dreamlike notions, with their time-honored
> traditions, suddenly assumed a terrifying timeliness. . . .
> The idea of humanity, purged of all sentimentality, has
> the very serious consequence that in one form or another
> men must assume responsibility for all crimes committed
> by all men, and that eventually all nations will be forced
> to answer for the evil committed by all others. Tribalism
> and racism are the very realistic, if very destructive, ways
> of escaping this predicament of common responsibility.
>
> Hannah Arendt, *The Origins of Totalitarianism*

At the end of the "Preface to First Edition" of *Origins,* Arendt writes, "Anti-Semitism (and not merely the hatred of Jews), imperialism (not merely conquest), totalitarianism (not merely dictatorship)—one after the other, one more brutally than the other, have demonstrated that human dignity needs a new guarantee which can be found only in a new political principle, in a new law on earth, whose validity this time must comprehend the whole of humanity while its power must remain strictly limited, rooted in and controlled by newly defined territorial entities" (*OT,* xi). At the very beginning of her seminal work, Arendt calls for a universal principle of humanity that will provide a new guarantee of human dignity. Arendt gives her reason for the need for such a principle at the conclusion of *Origins,* in her analysis of totalitarianism and the unprecedented reality of the death camps. I quote her at length:

In comparison with the insane end-result—concentration-camp society—the process by which men are prepared for this end, and the methods by which individuals are adapted to these conditions, are transparent and logical. The insane mass manufacture of corpses is preceded by the historically and politically intelligible preparation of living corpses. The impetus, and what is more important, the silent consent to such unprecedented conditions are the products of those events which in a period of political disintegration suddenly and unexpectedly made hundreds of thousands of human beings homeless, stateless, outlawed, and unwanted, while millions of human beings were made economically superfluous and socially burdensome by unemployment. This in turn could only happen because the Rights of Man, which had never been philosophically established but merely formulated, which had never been politically secured but merely proclaimed, have, in their traditional form, lost all validity. (446)

It is clear that Arendt places the responsibility for the death camps squarely at the feet of a philosophically invalid and politically impotent notion of human rights. This is not to suggest that she sees a causal link between the modern formulation of human rights and the event of totalitarianism. Instead, as she states in her well-known response to Eric Voegelin, she is tracing the elements that crystallized into totalitarianism rather than writing a history of totalitarianism as such.[1] But, she insists, we are able to see through the crystal to the ground in which it is embedded. For Arendt, the ground that at least in part provided the condition for this crystallizing event is the modern Declaration of the Rights of Man. Here it is worth noting that Arendt is not party to those who think it enough to simply repair this declaration in order to prevent the worst from happening again. Indeed, she ends the preface by stating that totalitarianism reveals a subterranean realm that renders all such rehabilitation projects futile: "The subterranean stream of Western history has finally come to the surface and usurped the dignity of our tradition. This is the reality in which we live" (ix). The crystallizing event of totalitarianism allows us a glimpse into the subterranean realm, revealing that the modern nation-state with its declaration of human rights is deeply entangled in a racism and an imperialism that call for something more than a restoration of the Enlightenment project.

Establishing philosophically and securing politically human rights requires a new law of humanity. In *The Origins of Totalitarianism*, Arendt argues:

Man of the twentieth century has become just as emancipated from nature as eighteenth-century man was from history. History and nature have become equally alien to us, namely, in the sense that the essence of man can no longer be comprehended in terms of either category. On the other hand, humanity, which for the eighteenth century, in Kantian terminology was no more than a regulative idea, has today become an inescapable fact. This new situation in which "humanity" has in effect assumed the role formerly ascribed to nature or history would mean in this context that the right to have rights or the right of every individual to belong to humanity should be guaranteed by humanity itself. It is by no means certain whether it is possible. (298)

The gates of Heaven are shut, the hands of God are closed. The rationality of nature, the self-evidence of reason, and the progress of history have given way to the death camps and holes of oblivion, leaving us to face nothing but ourselves. Humanity itself must guarantee the right to have rights, or the right of every individual to belong to humanity.

HUMANITY AND THE
QUESTION OF IDOLATRY

While Arendt appeals to humanity to guarantee the right to have rights, she is by no means idolatrous when making this appeal. Indeed, for her, the ideal of humanity is terrifying. In her analysis of the racism inherent in the imperialistic pan-movements, she argues against the popular cliché that the more we know about each other, the more we will like each other. On the contrary, Arendt argues, "The more peoples know about one another, the less they want to recognize other peoples as their equals, the more they recoil from the ideal of humanity" (235). In "Organized Guilt and Collective Responsibility," Arendt makes a similar claim. Pointing out that the modern world is characterized by increased knowledge of other cultures and peoples, she argues, "Since then peoples have learned to know one another better and learned more and more about the evil potentialities in men. The result is that they have recoiled more and more from the idea of humanity and become more susceptible to the doctrine of race, which denies the very possibility of a common humanity" (*EU*, 131). The political difficulty today, she argues, is that "idealistic talk about mankind and the dignity of man [is] an affair of the past simply because all these fine and dreamlike notions, with their time-honored traditions, suddenly assumed a terrifying timeliness" (*OT*, 235). Arendt does not make the claim that we cannot get along; instead, she claims that the element that unites us, humanity, is also the element that causes terror. With the deepening of our knowledge of others, we recoil all the more from the ideal of humanity. That ideal, when purged of all sentimentality, demands that humanity assume political responsibility for all crimes and evils committed by human beings. At the same time, this demand is terrifying. Herein lies the predicament of common responsibility.[2]

In her first extended published writing on political responsibility and the solidarity of humanity, Arendt elaborates on this predicament of common responsibility:

> For many years I have met Germans who declare that they are ashamed to be German. I have often been tempted to answer that I am ashamed to be a human being. This elemental shame, which many people of the most various nationalities share with one another today, is what finally is left of our sense of international solidarity; and it has not yet found an adequate political expression. (*EU*, 131)

Arendt argues that this sense of shame is the nonpolitical expression of the insight that "in one form or another men must assume responsibility for all crimes committed by human beings and that all nations share the onus of evil committed by all" (131). The international solidarity of humanity lies in this almost-unbearable burden of global political responsibility; it is a solidarity rooted in facing up to the human capacity for evil:

> Those who today are ready to follow this road in a modern version do not content themselves with the hypocritical confession, "God be thanked, I am not like that" . . . in horror of the undreamed of potentialities of the German national character. Rather, in fear and trembling, have they finally realized what man is capable of—and this indeed is the condition for any modern political thinking. (132)

In her 1954 essay "Concern with Politics in Recent European Thought," Arendt does not change her mind concerning the condition for modern political theory. While agreeing with the Greeks that philosophy begins with wonder at what is, Arendt harbors no nostalgia for the Greek experience. Instead, she claims that whereas the Greek experience of wonder was rooted in the experience of beauty (*kalon*), the experience of wonder today —if not engaged in a flight from reality—is rooted in the experience of horror at what humans are capable of, the speechless horror that philosophically must be endured and politically instituted against:

> It is as though in this refusal to own up to the experience of horror and take it seriously the philosophers have inherited the traditional refusal to grant the realm of human affairs that *thaumadzein*, that wonder at what is as it is. . . . For the speechless horror at what man may do and what the world may become is in many ways related to the speechless wonder of gratitude from which the questions of philosophy spring. (445)

Speechless horror, not beauty, marks the contemporary experience of wonder. This facing up to the human capacity for evil separates Arendt from
her Enlightenment predecessors, who, she argues, were too naïve in their
view of humanity: "Our fathers' enchantment with humanity was of a sort
which not only light-mindedly ignored the national question; what is far
worse, it did not even conceive of the terror of the idea of humanity and
of the Judeo-Christian faith in the unitary origin of the human race" (132).
Here again Arendt points to the terror inherent in the idea of humanity,
linking that idea to the notion of a *unitary* origin of the human race. As
we shall see, Arendt's location of the origin in the originary event of natality not only criticizes this notion but offers the possibility of lessening the
terror.

Arendt is not at all enchanted, but at the same time she refuses to simply abandon the idea of humanity. She argues that such abandonment is
impossible insofar as the idea of humanity, "which for all preceding generations was no more than a concept or an ideal, has become something of
an urgent reality" (*MDT,* 82). In "Organized Guilt and Responsibility," she
goes farther, arguing for its political necessity: "In political terms, the idea
of humanity, excluding no people and assigning a monopoly of guilt to no
one, is the only guarantee that one 'superior race' after another may not feel
obligated to follow the 'natural law' of the right, of the powerful, and exterminate 'inferior races'" (*EU,* 131). Only a principle of humanity is able
to provide the normative source for an imperative of common responsibility. For all the horror at the very heart of human relations, and despite her
rejection of a metaphysical notion of human nature, Arendt remains a hu-
manist. And as an earlier passage makes clear, for Arendt, humanity itself
most now assume the role formerly ascribed to nature, history, or god:
"The right to have rights, the right of every individual to belong to humanity, should be guaranteed by humanity itself" (*OT,* 298). Against the
Enlightenment, Arendt disavows the goodness of human nature, insisting
on our very real capacity for evil. Against Ignatieff, Arendt claims that when
it comes to articulating humanity, we can move beyond faith without falling into idolatry.

For Arendt, humanity's guarantee lies not in the end of humanity but
in its beginning. In *The Human Condition,* Arendt restates this claim:

> To act, in its most general sense, means to take an initiative, to begin (as the
> Greek word *archein,* "to begin," "to lead," and eventually "to rule" indicates),
> to set something into motion (which is the original meaning of the Latin
> *agere*). Because they are *initium,* newcomers and beginners by virtue of
> birth, men take initiative, are propelled into action. *[Initium] ergo ut esset,*
> *creatus est homo ante quem nullus fuit* ("that there be a beginning, man was

created before whom there was nobody"), said Augustine in his political philosophy. This beginning is not the same as the beginning of the world; it is not the beginning of something but of somebody, who is a beginner himself. With the creation of man, the principle of beginning came into the world itself, which, of course is only another way of saying that the principle of freedom was created when man was created but not before. (*HC,* 177)

For Arendt, the event of natality is the *arche* in the double etymological sense of origin and rule. Further, the unpredictable anarchic origin carries its rule or principle within it. As she points out, "What saves the act of beginning from its own arbitrariness is that it carries its own principle within itself or, to be more precise, that beginning and principle, *principium* and principle, are not only related to each other, but are coeval." Arendt goes on: "For the Greek word for beginning is *arche,* and *arche* means both beginning and principle" (*OR,* 212). Arendt locates the principle of humanity that grounds the right to have rights in this archaic event of natality. As we will see, the principle (*arche*) of this event is double: the principle of publicness and the principle of givenness.

Before proceeding to an examination of the archaic event of natality, we must consider a further objection to this endeavor to articulate an ontological foundation of human rights. The objection, formulated by thinkers such as Michael Ignatieff, John Rawls, and Claude Lefort, argues that any attempt to ground human rights is unnecessary at best and dangerous at worst. We have already seen that Ignatieff articulates this objection when he states that if we avoid idolatry, "human rights are nothing other than a politics."[3] Human rights for Ignatieff do not proclaim "eternal verities" but instead "create a common framework, a common set of reference points that can assist parties in conflict to deliberate together."[4] Human rights must be seen in terms of a political discourse that allows for "the adjudication of conflict." At the same time, Ignatieff wants to claim more, arguing that "another essential function of international human right covenants, even in societies with well-ordered rights regimes, is to provide a universalist vantage point from which to criticize and revise particularistic national law."[5] The problem is that having reduced human rights to history and politics, Ignatieff has no way to flesh out this "universalist vantage point."

Like Ignatieff, Rawls also argues that we do not need an ontological foundation for human rights because human rights are nothing more than political rights. In *The Law of Peoples,* Rawls argues that the category of the political provides all we need:

A third condition for a realistic utopia requires that the category the political must contain within itself all the essential elements for a political concept of justice. For example, in political liberalism, persons are viewed as citizens and a political conception of justice built up from political (moral) ideas is available in the public political culture of a liberal constitutional regime. The idea of a free citizen is determined by a liberal political conception and not by any comprehensive doctrine, which extends beyond the category of the political.[6]

Rejecting "any comprehensive doctrine," by which I take him to mean any kind of ontology that extends beyond the category of the political itself, Rawls argues that these political rights emerge from "the public political culture of a liberal constitutional regime." The problem of reducing human rights to a political principle of liberalism is that this again equates rights with the status of the citizen. Rawls himself seems to recognize that in the case of human rights, this equation may not be sufficient. He argues that human rights "express a special class of urgent rights, such as freedom from slavery and serfdom, liberty (but not equal liberty) of conscience, and security of ethnic groups from mass murder and genocide."[7] In the context of these urgent rights, Rawls wants to maintain that human rights are something *other* than politically instituted positive rights: "Human rights are distinct from constitutional rights, or from the rights of liberal democratic citizenship, or from other rights that belong to certain kinds of political institutions, both individualist and associationist."[8] However, he goes on to argue that human rights are "a proper subset of the rights possessed by citizens in a liberal constitutional democratic regime, or of the rights of the members of a decent hierarchical society."[9] With this, Rawls puts human rights squarely back in the context of citizenship. Thus, he argues that outlaw states that violate these rights are to be condemned. But on the basis of what right? His answer: on the basis of the liberal constitutional democratic regime, which presumably determines who is an "outlaw."

Rawls's rejection of ontology (again, what he calls "a comprehensive philosophical principle") costs him more than it benefits. Searching for a way to avoid historicism, Rawls ends up unable to escape it. Rights belong to those lucky enough to have landed historically in a liberal constitutional state. Those states who deny human rights are condemned on the basis of the validity of the liberal state, whose validity, in turn, is assumed on the basis of its mere existence.[10] Habermas calls this a "defiant appeal to the factual"[11] and is in agreement with Arendt that such an appeal is not sufficient to establish the universality of human rights, a universality that both Ignatieff and Rawls want to posit: "The development of constitutional democracy along the celebrated 'North Atlantic' path has certainly provided us with results worth preserving, but once those who do not have the good fortune to be heirs of the Founding Fathers turn to their own traditions, they cannot find criteria and reasons that would allow them to distinguish what is worth preserving from what should be rejected."[12]

Habermas's charge of a "defiant appeal to the factual" can also be leveled against Claude Lefort. Lefort refers to Arendt's right to have rights to argue in favor of human rights against those who would point to the uselessness of invoking human rights in societies where a large part of the population is "a victim of savage exploitation."[13] He argues that "experience teaches us only too clearly that scorn for human rights encourages would-be revolutionaries to construct totalitarian-style regimes, or to dream of doing so. It masks an underlying refusal to grant individuals, peasant communities, workers, and peoples in general *the right to have rights*."[14] For Lefort, however, the instituting principle of democracy, articulated in the declaration of human rights, is a principle of contestation wherein the question of right is "always dependent upon a debate as to its foundations, and as to the legitimacy of what has been established and of what ought to be established."[15] More precisely, the democratic contest is instituted through the declaration of right that opens an irreducible space between the sphere of power, the sphere of law, and the sphere of knowledge. At the same time, Lefort argues that no universal basis can be given to human rights: "The dimension of the development of right unfolds in its entirety, and it is always dependent upon a debate as to its foundations, and as to the legitimacy of what has been established and of what ought to be established."[16] For Lefort, the notion of right in a democracy "depends upon the discourse which articulates it, and in which the exercise of power depends upon conflict."[17] Thus, Lefort denies any universal basis for human rights: "Fundamental rights may well be constitutive of a public debate, but they cannot be constrained by a definition; and we therefore cannot agree on any universal basis as to what conforms or does not conform to the letter or the spirit of those rights."[18] Human rights are for Lefort always dependent upon a specific political debate for their foundation.

Similar to Rawls in this regard, Lefort's analysis works in the context of modern democracies that have been constituted by a declaration of right. But what of those political spaces where no declaration has taken place? What of those political places where no debate about right is possible because no claim to right has been recognized as valid? To argue that it is up to individuals to claim rights through a debate is to miss the political urgency: those who desperately appeal to human rights are often those who are in no position to be recognized as claimants before a tribunal that has already decided against them. Thus, from an Arendtian point of view, we must go beyond the debate about what is legitimate and illegitimate and provide a universal basis for the right to have rights. Only from this universal framework is it possible to delineate the legitimate and illegitimate uses of power.

THE ARCHAIC EVENT OF NATALITY

Arendt is very explicit that the event of natality is an ontological event. In *The Human Condition,* she writes: "The miracle that saves the world, the realm of human affairs, from its normal 'natural' ruin is ultimately the fact of natality, in which the faculty of action is *ontologically rooted*" (*HC,* 246, emphasis mine). At the same time, she is equally insistent that this ontological event is not metaphysical; it is not the origin of anything like human nature: "To avoid misunderstanding, the human condition is not the same as human nature and the sum total of human activities and capabilities which correspond to the human condition do not constitute anything like human nature" (9). Indeed, this event has the character of a "startling unexpectedness." Natality, she argues, is the condition for human existence, but it can never "explain what we are or answer the question of who we are for the simple reason that [it] can never condition us absolutely" (10). The "who" does not possess an enduring fixed nature but is instead inherently marked by contingency and unpredictability. Arendt's ontology, therefore, does not describe an immutable order of essences. It does not seek enduring truths upon which to ground both thought and action; it does not posit a metaphysical notion of human nature or subjectivity in which human rights are inalienably inscribed. Instead, it is rooted in an event that provides the *arche* and *principium* of human action. By articulating this *principium,* Arendt does not give us an ontological politics; rather, she provides an ontological foundation for human rights.

Recognizing Arendt's debt to Montesquieu is essential if we are to grasp the ontological status of the *principium* at work in the event of natality, especially in clarifying the distinction I am claiming between an ontological foundation of the political and an ontological politics. Arendt is most interested in Montesquieu's discovery that "each form of government has its own innate principle which sets it into motion and guides all its actions" (*EU,* 331). As is well known, Montesquieu argues in *Spirit of the Laws* that the *form* of a government is always animated by a *spirit* or *ethos* that animates the various institutions and laws of a particular form of government. He understands the *spirit* or *ethos* to be an affection or passion that provides the principle of action within a particular regime: "There is this difference between the nature and principle of government that the former is that by which it is constituted, the latter that by which it is made to act. One is its particular structure, and the other the human passions which set it in motion."[19] Thus, Montesquieu argues that a monarchy is animated by love of honor, a republic by love of virtue, and a tyranny by fear. The animating affection is the origin (*arche*) of action and as such carries its rule or principle within it.

What interests Arendt is Montesquieu's claim that political principles are different from the laws that order a particular political space. Laws, she argues, establish limits or boundaries that circumscribe and stabilize action: "The law defines the boundaries of personal life but cannot touch what goes on within them. In this respect, the law fulfills two functions: it regulates the public-political sphere in which men act in concert as equals and here they have a common destiny, while, at the same time, it circumscribes the space in which our individual destinies unfold" (334). Principles, on the other hand, are sources of action and motion, providing the "common ground in which the laws are rooted and from which the actions of citizens spring." Principles are "moving principles"; they orient action and "map out certain directions." Describing these principles, she states: "Hedged in by law and power, and occasionally overwhelming them, lie the origin of motion and action" (335).

Most important for grasping the *principium* inherent in the event of natality is Arendt's claim that these principles of movement and action are at work in both the public and private realms. Indeed, the principles of movement and action unify the public and private, thereby resolving for Arendt one of the most difficult problems in political thought; namely, "the discrepancy between public and personal life, between man as citizen and man as individual." The discrepancy, she argues, cannot be resolved by the law, which "can never be used to guide and judge actions in personal life" (334). Indeed, from the perspective of the law, these two spheres are often in conflict. Seemingly at odds with what she argues in *The Human Condition*, Arendt claims that we can discover the unity between the public and the private in these principles of action. I quote "On the Nature of Totalitarianism" at length:

> Since there was an obvious, historically patent correspondence between the principle of honor and the structure of monarchy, between virtue and republicanism, and between fear (understood not as a psychological emotion but as a principle of action) and tyranny, then there must be some underlying ground from which both man as an individual and man as a citizen spring. In other words, Montesquieu found that there was more to the dilemma of the personal and the public spheres than discrepancy and conflict, even though they might conflict. (335)

Two characteristics of these principles need to be underscored. First, principles of action have an affective dimension that are not reducible to psychological emotion but belong to the principle of action itself. These affective principles of action orient human emotions, inspiring them to move in one direction rather than another. Arendt states: "The common ground of republican law and action is thus the insight that human power is not primarily limited by some superior power, God or Nature, but by the powers of one's equals. And the joy that springs from that insight, the 'love of equality' which is virtue, comes from the experience that only because this is so, only because there is equality of power, is man not alone" (336). Thus, the affectivity, in this case the love of virtue that carries with it the joy of equality, is inherent in the principle itself. Still further, these affective principles provide a normative orientation to action: "Even in the personal sphere, where no universal laws can ever determine unequivocally what is right and what is wrong, man's actions are not completely arbitrary. Here he is guided not by laws, under which cases can be subsumed, but by principles—such as loyalty, honor, virtue, faith—which, as it were, map out certain directions" (335). Orienting us affectively, these principles provide the normative source of all types of action, both public and private, in each form of government.

This leads to the second characteristic of the *principium* at work in the event of natality. Principles of action provide the affective ground of unity between the individual and the citizen. Arendt argues: "The phenomenon of correspondence between the different spheres of life and the miracle of the unities of cultures and periods despite discrepancies and contingencies indicates that at the bottom of each cultural or historical entity lies a common ground which is both fundament and source, basis and origin" (335). The strict distinction she makes between the public and the private in the later *The Human Condition* is in fact more nuanced in her earlier work. In her reading of Montesquieu, she claims that while there is a distinction between the private and the public, they are unified in the common source provided by the inspiring affective principle that orients action in both domains.

In her later essay "What Is Freedom?" Arendt further develops her notion of a principle of action that provides common ground between the private and the public, the individual and the citizen. Referring again to Montesquieu, she argues that principles of action are not inherent in the self, nor do they prescribe particular goals: "Principles do not operate from within the self as motives do—but inspire, as it were, from without, and they are much too general to prescribe particular goals, although every particular aim can be judged in the light of its principle once the act has been started." Principles of action "map out certain directions"; they orient action without prescribing it. Arendt insists that principles must be enacted and "become fully manifest in the performing act itself." In other words, publicity is essential to the principles; they can be seen in the act itself.

Speaking specifically about the principle of freedom, she writes, "Freedom or its opposite appears in the world whenever such principles are actualized; the appearance of freedom, like the manifestation of principles, coincides with the performing act" (*BPF,* 151–153).

This means that principles are neither concrete nor abstract; they do not prescribe anything in particular, while at the same time they are always on view in the action itself. Moreover, unlike motives, which are always particular, "the validity of a principle is always *universal* and is not bound to any particular person or to any particular group" (emphasis mine). Through the activity itself, a principle of action is always embedded in history, while at the same time it transcends any particular historical moment or action. Arendt argues that "the principle of action can be repeated time and again, it is inexhaustible" (*BPF,* 152). Here we must be cautious. Arendt is not claiming that principles are atemporal or eternal. Her claim that principles need to be enacted in order to be fully manifest suggests that a temporal dimension is always present in the manifestation of a principle. The principle's inexhaustibility, on the contrary, describes the way in which it can never be fully realizable in any particular concrete action. In other words, the principle's inexhaustibility renders futile any and all attempts to realize it once and for all in a particular political program or agenda.

It is this last aspect of a political principle that especially underpins Arendt's insistence on the distinction between *poiesis* and *praxis,* between the "in-order-to" and the "for-the-sake-of." Activities of *poiesis* (making) are guided by principles that can be fully inscribed in what results from the activity. For example, principles of good craftsmanship can be used to make a table; these principles are inscribed in the completed table and remain so after the activity of making is finished. This is not the case with principles of action. The inexhaustibility of these principles entails that they can never be inscribed in any concrete action. Principles of action inspire. They move us to act for the sake of the principle. They are manifest in the act as long as the activity endures, but no longer. Thus, they can never be realized in the same way that a table can be realized and be put to good use in the family dining room for decades to come. The principle of justice only moves us in the direction of just actions; it only orients us in our pursuit of justice. It does not allow for the realization of justice on earth.[20]

This last is especially important for understanding why Arendt is not providing an "ontological politics," if by that is meant a politics that finds its firm ground in a set of enduring truths that can then be realized in the political realm. The ontological foundation of the political provides *only* the universal principle that inspires and orients political action; it neither provides a model for political action nor prescribes any particular political actions. Further, the inexhaustibility of the principle entails that it always transcends the concrete political space. An irreducible space opens up between the ontological and the political. For Arendt, an irreparable divide and difference separates the two realms, rendering impossible the reduction of the political to the ontological. As in the case of thinking, we must "act without a banister"; that is, we must act without the support of an ontology that would tell us what to do.

Arendt concludes her discussion of Montesquieu in "On the Nature of Totalitarianism" by considering the source of these principles. While the principles of honor, virtue, and fear are the inspiring principles of monarchies, republics, and tyrannies, respectively, what are the sources of the principles themselves? Arendt answers: "These three forms of government—monarchy, republicanism, and tyranny—are authentic because the grounds on which their structures are built (the distinction of each, equality of all, and impotence) and from their principles of motion spring are authentic elements of the human condition and are reflected in primary human experiences" (*EU,* 338). The source of these principles, she argues, is the human condition itself. For Arendt, the human condition is characterized by the distinctness and uniqueness of each individual (reflected in the love for distinction), the plurality within which this uniqueness always already finds itself (reflected in the love of equality), and the loneliness that ensues when plurality is replaced with radical isolation (reflected in impotent fear and the will to dominate). Arendt concludes by reflecting on totalitarianism as an "unprecedented form of government," asking whether this new form of government "can lay claim to an equally authentic, albeit until now hidden, ground of the human condition on earth, a ground which may reveal itself only under circumstances of a global unity of humanity—circumstances certainly as unprecedented as totalitarianism itself" (338). Here Arendt points to a new, until now hidden ground of the human condition that is the source of a principle of humanity coincident with the unprecedented condition of global unity. (Totalitarianism, she argues, lays bare the unprecedented danger that ensues in a condition of global unity when this principle is not the inspiring source of political action.) This source or ground of a principle of humanity is the ontological event of natality.

At this point an objection may be raised. Is it not the case that Arendt consistently argues against reducing the political to the natural? Does she not time and again insist that "the political is not the natural"?[21] And does not the attempt to locate the foundation of human rights in the event of natality do precisely that; namely, ground the political on the most natural physical event of all, the event of human birth? Georgio Agamben makes such an objection in his reading of Arendt, arguing that her emphasis on

the event of natality as the ground of political action brings her dangerously close to the precipice that she herself wishes to avoid. If we ground
the political generally and human rights specifically in a natural and physical event, he asks, is this not to court the danger of biologism all over
again? Agamben suggests that the danger facing the political today is not so
much metaphysics, with its various notions of human nature, as it is these
biological organic fantasies that seek to provide grounds for the many
racist and ethnic ideologies that the twentieth and twenty-first centuries
know all too well.[22] Responding to Agamben's objection, I argue in the
next section that the temporality of this event is such that natality is never
merely a physical or natural event, even though it does have an inherent dimension of facticity. Here Arendt's discussions of Kafka and Benjamin are
most important. I then argue in section four that despite Arendt's own
claim that the event of natality is *at first* the "naked fact of physical birth,"
this claim does not hold up when considering the several places in her
work where she suggests that this physical birth is *from the outset* inseparable from a "linguistic birth" that renders it impossible to reduce the event
of natality to a physical biological event. Here her engagement with Heidegger is crucial.

THE TEMPORALITY OF NATALITY

The importance of Kafka's parable "He" to Arendt's thought cannot be un-
derestimated. In both *The Life of the Mind, Thinking* as well as the preface
to *Between Past and Future,* she discusses this parable at length. In a footnote to the title of her essay "Truth and Politics" (in *Between Past and Future*), Arendt again points to this parable as a way to understand the temporality implicitly at work in her judgments of the Eichmann trial. And in
Essays in Understanding, she devotes an entire essay to "The No Longer and
the Not Yet," invoking once again the temporality at work in Kafka's parable. Arendt's recognition that two dimensions are present in this tale is important to our discussion. On the one hand, it is describing the temporality of judging and acting, and on the other, it is describing the temporality
of natality itself.

Briefly, in this parable the contemporary thinker, "He," stands in the
gap between the past and the future. "He" is engaged in a battle with two
antagonistic forces, one coming from the infinite past and the other from
the infinite future. The past is driving him into the future, and the future
is driving him back to the past. For Arendt, this describes the condition of
thought that has always existed. However, in other times, this gap was
paved over by tradition. It is only with the loss of tradition that the gap has
revealed itself in its true character, engaging the thinker in the difficult

struggle that Kafka describes. "He" would like to leap beyond this struggle and assume the role of umpire. For Arendt, this is the metaphysical dream: to transcend the finitude of existence and enter a silent realm of the eternal. At this point, Arendt alters the parable. I quote her text at length:

> Without distorting Kafka's meaning, I think one may go a step further. Kafka describes how the insertion of man breaks up the unidirectional flow of time but, strangely enough, he does not change the traditional image according to which we think of time as moving in a straight line. . . . The trouble with Kafka's story in all its magnificence is that it is hardly possible to retain the notion of a rectilinear temporal movement if its unidirectional flow is broken up into antagonistic forces being directed toward and acting upon. (*BPF* 11)

Arendt argues that the battle between the past and the future produces a "*deflection of forces,*" and, further, that this deflection produces a third force that is diagonal to the forces of the past and the future: "Ideally, the action of the two forces which form the parallelogram of forces where Kafka's 'he' has found his battlefield should result in a third force, the resultant diagonal whose origin would be the point at which the forces clash and upon which they act" (12; my italics).

Three points must be noted. First, the origin of the diagonal force is antagonistic and it arises only in the interstitial between the past and the future. Thus, it can be located between the past and the future but cannot be grounded in either. In other words, it is a temporal origin that is deflected from any foundation in the past. It is not an origin that will ground the future, either.

Second, the gap between the past and future is not created through the break in tradition. Rather, tradition is that which has covered over this gap, making it easier to move from the past to the future. Arendt argues that tradition paves the road between the past and the future, thereby allowing thought and action to move along a straight and unimpeded path. It is tradition that gives the sense that the movement between past and future is linear. Arendt argues that the gap is created because human beings are beginners who by birth are inserted between the force of the past and the force of the future, and as beginners they continually disrupt the linear form of time: "The insertion of man, as he breaks up the continuum, cannot but cause the forces to deflect, however lightly, from their original direction, and if this were the case, they would no longer clash head on but meet at an angle" (11). Arendt's analysis of temporality in the Kafka parable, therefore, is an analysis of the temporality of natality itself, and not, as it has so often been read, an analysis of a kind of modern temporality that has occurred through a break in tradition. Most important for our discussion, insofar as the gap between the past and the future is created by the insertion of beginners, Arendt's analysis extends beyond a consideration of the temporality of thinking and judging to include the temporality of action and, by extension, the temporality of natality itself, which she defines as the insertion in linear time of unique beginners.

Third, as noted above, the gap between the past and the future is a "moment of deflection" that produces a third force. This is the force of the present. The present is neither a complete break with the past (a kind of rupture) nor is it a moment of transition to the future. Instead, the force of the past, rather than being originary, is deflected, thereby projecting something aberrant or other into the present. The future, the "not-yet," emerges in the deflective force of the past. There will have been movement back (or, as we shall see, *backward*) to the future, another way to think the future anterior. Arendt's reading of the Kafka parable suggests that the past is an anteriority that constantly introduces an aberration or a difference into the future through this deflective or disjunctive present. The deflective present, therefore, does not simply repeat the past but also initiates the new.

Arendt's essay on Benjamin illuminates her reading of the Kafka parable, specifically the temporality of the deflected present, which marks the temporality of the event of natality. Arendt begins the essay by noting Benjamin's awareness of the break in tradition and the loss of authority:

> Insofar as the past has been transmitted as tradition, it possesses authority; insofar as authority presents itself historically, it becomes tradition. Walter Benjamin knew that the break in tradition and the loss of authority which occurred in his life-time were irreparable, and he concluded that he had to discover new ways of dealing with the past. In this he became a master when he discovered that the transmissibility of the past has been replaced by its *citability* and that in place of its authority there had arisen a strange power to settle down, piecemeal, in the present and to deprive it of "peace of mind," the mindless peace of complacency. (*MDT,* 193, italics mine)

The untransmissibility of the past makes the present unsettling. Moreover, as in her reading of the Kafka parable, Arendt does not understand this break in the tradition as a kind of rupture with the past. Instead, she gives close attention to how Benjamin's "angel of history" and his notion of "citability" allow for a different understanding of history and, further, how this notion of history contributes to the emergence of the unprecedented and the unique. This, in turn, allows us to better grasp the emergence of the new in the event of natality.

Arendt compares Benjamin's angel of history with his description of the *flaneur:* "For just as the *flaneur,* through the *gestus* of purposeless strolling, turns his back to the crowd even as he is propelled and swept by it, so the 'angel of history,' who looks at nothing but the expanse of ruins in the past is blown backwards into the future by the storm of progress" (165). Arendt suggests that Benjamin's angel of history is not merely offering a critique of the notion of historical progress but also is at the same time articulating a notion of a future anteriority. Benjamin's angel of history moves back(ward) to the future. Arendt emphasizes how this future anteriority introduces the foreign and the new. In the context of her analysis of Benjamin's claim that there is no simple return to the tradition, she writes, "It was an implicit admission that the past spoke directly only through things that had not been handed down, whose seeming closeness to the present was thus due precisely to their exotic character, which rules out all claims to a binding character" (195). Arendt points out that for Benjamin there could be no return to a tradition that could have binding force. Reflecting on Benjamin's interest in the baroque, which, Arendt states, "has an exact counterpart in Scholem's strange decision to approach Judaism via the Cabala, that is, that part of Hebrew literature which is *untransmitted and untransmissible* in terms of Jewish tradition," she argues that nothing so clearly showed "that there was no such thing as a 'return' either to the German or the European or the Jewish tradition than the choice of these fields of study" (195, italics mine). Instead, the angel of history moves into the future by citing that which is untransmitted in terms of the tradition.

The dual notions of citability and translation are critical to Arendt's reading of Benjamin's understanding of the angel of history. Arendt points to Benjamin's reflections on the use of quotations: "Quotations in my works are like robbers by the roadside who make an armed attack and relieve an idler of his convictions" (193). Citability is a force that destroys the complacency of the present by robbing from the past that which is foreign and unfamiliar to the present. Moreover, it has a "destructive power" that nonetheless "still contains the hope that something from this period will survive—for no other reason than that something was torn out of it" (193). Citation invokes the past with a deadly impact directed against tradition and its authority. Thus, "the heir and preserver unexpectedly turns into a destroyer" (199). It is, however, a sacrifice and a destruction that opens the future to the new: "The genuine picture may be old, but the genuine thought is new. It is of the present. This present may be meager, granted. But no matter what it is like, one must firmly take it by the horns to be able to consult the past. It is the bull whose blood must fill the pit if the shades of the departed are to appear at its edge" (199).

Arendt goes on to say that quotations accomplish this by "interrupting the flow of the presentation with 'transcendent force' and at the same time of concentrating within themselves that which is presented" (194). The "transcendent force of citability" is another way of describing that force of deflection Arendt evokes in her rewriting of Kafka's parable. The transcendent force of citability makes the present a moment of deflection whereby the past becomes a projective force of the unknown and the unfamiliar; that is, a projective force of the new. Citability, therefore, offers a notion of historical narrative that is not descriptive but inaugurative. Its inaugurative force is derived precisely from its decontextualization, from its break with a prior content and its capacity to assume new contexts; it introduces a reality rather than reports on what already exists. And the paradox is that it accomplishes this introduction through a citation of what already exists. Thus, Arendt argues that the pearl diver (one who is engaged in this activity of citability) is an alchemist and not a commentator. The activity of citability in the deflected present makes graphic a moment of transformation, not merely the continuum of history.

We need to recall that citability not only destroys the complacency of the present but also at the same time introduces from the past what is new and foreign. This new element, however, needs translation. In her reading of Benjamin's essay "The Task of the Translator," Arendt writes: "What mattered to him above all was to avoid anything that might be reminiscent of empathy, as though a given subject of investigation had a message in readiness which easily communicated itself, or could be communicated, to the reader or spectator" (203). Then Arendt quotes Benjamin's essay: "No poem is intended for the reader, no picture for the beholder, no symphony for the listener" (203). There is no communication of the given, no interpretation of what is already there. The pearl diver has the task of translation, not hermeneutics—there is no reflection on an already given content or object. Instead, translation is engaged with what is not given, something foreign that cannot be empathized with, something incommensurable that cannot be reduced to the same, an unknown that cannot be reduced to the known.

Here we need to pause and consider why Benjamin's work on translation is so important for Arendt's notion of natality. As we have seen in the above remarks on the angel of history, the new is not part of a progressivist discourse. The newness is not contained in any commentary on some pre-given reality (hermeneutics) but instead requires an act of alchemy that occurs through citability of what is untransmitted and untransmittable (translation). The untransmittable, in particular, reveals why citability requires translation. Citability introduces foreignness in the double sense of the untransmitted and the untransmittable. Thus, the foreign element is never overcome through the activity of citability. In other words, the incommensurable element is never surmounted; it can never be altogether assimilated into the present narrative.[23]

This last is tremendously important for understanding the temporality at work in the event of natality. In the context of a discussion concerning the "who" and its insertion into the web of relationships, Arendt takes up the question of the temporality of action. What is temporal in the mode of the existence of the political actor? Arendt answers that the reification of the actor can occur only through a "kind of repetition, the imitation of mimesis, which according to Aristotle prevails in all arts but is actually appropriate to the drama" (*HC*, 187). Arendt's qualification of the "kind of repetition and mimesis at work in deed and speech" is important because it indicates that she is distancing herself from a simple understanding of repetition or mimesis. In Aristotle (who is quite different from Plato in this respect), mimesis allows reality to appear in a new way. Mimesis adds something to reality, and it seems to me that this what Arendt has in mind in her reference to Aristotle in *The Human Condition*. For Aristotle, mimetic production or representation is not reproductive but inscribes differently. This inscription of a difference is closely interwoven with time and repetition.

That she is thinking Aristotle's notion of repetition and mimesis is clear from our analysis of her reading of Kafka and Benjamin. Recall that in her reading of the Kafka parable, the temporality at work in judgment and action does not simply repeat the past. Instead, the force of the past antagonistically meets the force of the future, giving rise to a deflected present. Moreover, recall that Arendt's analysis of the parable specifically claims that the origin of judgment and action occurs in this deflective present that characterizes the gap between the past and the future. Thus, the origin of action—that is, speech and deed—does not have its ground in any notion of a progressive past that could simply be repeated. That this is the case allows for the emergence of the new. Thus, the so-called origin of action as the gap between past and future is always removed from any origin or ground as such. It is an origin that is abyssal. It is, as Derrida will much later say, the nonorigin of origins. Finally, recalling Arendt's reading of Benjamin, specifically the notion of citability, there is an infelicity in all enacted speech, an inherent independence from any of the specific contexts in which speech occurs. The force of enacted speech, then, is not merely determined from prior usage but issues forth precisely from its break with any and all prior usage. Thus, enacted speech is not descriptive but inaugurative—it introduces a novel reality rather than reports on an existing one. To use Derrida's distinction, the temporality of the deflective present is not a repetition that simply repeats the past. Instead, the temporality of the deflected present is iterative, thereby opening up new possibilities.[24] The repetition of iterability is interruptive and inaugurative—it is the temporality of the deflected present in which the actor seeks reinscription and revision of the web of relationships in which he or she is immersed. The process of reinscription and revision—the insertion or intervention of something that takes on new meaning—occurs in the temporal break of the deflected present.

Returning now to the consideration of the temporality characteristic of the event of natality, we can see that the origin or *arche* that allows for the unprecedented temporal insertion of a beginner remains itself an *anterior* origin. In other words, not only is the temporal insertion of the beginner characterized as a deflected present in relation to both the future and the past, the condition for this insertion is itself always deflected; it is a deflected beginning in the sense that there is no possible return to the origin. In this way, the beginner's relation to its origin is one of citability and translation of what remains untransmitted and untranslatable. Although in the event of natality no absolute rupture of the beginner from its beginning occurs (its principle continues to be in play throughout the life of the beginner), nonetheless the *arche* or origin itself is unrecoverable. Just as the unique beginner inserts the foreign into an already existing world, so too does the origin of this beginning remain foreign. Arendt is providing an ontological foundation to the political but not an ontological politics: the origin or ground of the political is not recoverable or accessible. This returns us to a point made earlier: all political fantasies premised on an unmediated return to the origin are chimeric.

PHYSICAL AND LINGUISTIC NATALITY

Flynn's criticism, however, has not been entirely addressed. While it may be granted that there is no return to the origin itself, the above analysis does not give an argument against understanding the event of natality as a natural physical event that, in turn, would allow for all types of organic and biological political fantasies to be developed, whether in translation or not. Indeed, Arendt herself seems to suggest that the event of natality is, at first, a physical event. In *The Human Condition,* she writes, "With word and deed we insert ourselves into the human world, and this insertion is like a second birth, in which we confirm and take upon ourselves the naked fact of our original physical appearance" (*HC,* 176). I submit, however, that Arendt is not entirely consistent in her claim that this event is originally one of naked physical appearance. As I suggested, her references to Heidegger are especially important for noting her own inconsistencies; they allow for a different understanding of the event of natality, one that shows the inseparability of the embodied dimension and the linguistic (symbolic) dimensions of this event.

Certainly Arendt does not waver from her claim that the birth of the "who" is a linguistic event: "The point is that the manifestation of the 'who' comes to pass in the same manner as the notoriously unreliable manifestations of ancient oracles, which, according to Heraclitus, 'neither reveal nor hide in words but give manifest signs'" (182). Language, she argues, allows human beings to appear, and without this linguistic birth, humans would literally be dead to the world: "A life without speech and without action . . . is literally death to the world; it has ceased to be a human life because it is no longer lived among men" (176). Through this *linguistic* birth, humans become political kinds of beings. Arendt cites Aristotle's definition of man as *zoon logon ekhon,* one for whom exists "a way of life in which speech and only speech made sense" (27). This linguistic birth is the birth of the "who"; that is, the birth of the unique self. Thus, the event of linguistic natality is the birth of the unexpected and the new. In other words, the birth of the political self is the birth of the unexpected word.

Insofar as Arendt insists that the unexpected and the new appears only in action, linguistic natality is inseparably linked to action. Indeed, Arendt never finds one without the other: "Without the accompaniment of speech, at any rate, action would not only lose its revelatory character, but, and by the same token, it would lose its subject. . . . Speechless action would no longer be action because there would no longer be an actor, and the actor, the doer of deeds, is possible only if he is at the same time the speaker of words" (178–179). By word and deed we insert ourselves into the public space. While Arendt is not entirely clear, she seems to suggest in this passage that speech has priority over action in the appearance of the "who." The self is not a consequence of speech; instead, the "who" is born in the very speaking itself. For Arendt, therefore, the unexpected word is performative; it is inaugurative and not descriptive. This explains why Arendt continuously refers to the theater when discussing the appearance of the self. Indeed, in *On Revolution,* she refers to the Greek notion of the self as *persona,* the voice that speaks through the mask in ancient tragedy. For Arendt, the important point is that there is no one behind the mask—the self is the *persona,* the voice that shines through.[25] The self as the unexpected word is the performance of the new. This self, therefore, is never representational or positional. Linguistic natality is the birth of the singular unique self, "before whom there was nobody" (177).

It would be easy, but also a mistake, to think of this second birth as the birth of a kind of heroic individuality, distinct in the sense of being "a word unto itself." Arendt, however, rejects any notion of the self as a "singular word," arguing that the unexpected word is always already immersed in a web of relationships and plurality of enacted stories (181). The Arendtian notion of a web reveals that the unexpected word erupts within a plurality of discourses that are entangled and interwoven in their sedimented histories. At the same time, and this is extremely important for our discussion of the event of natality, Arendt argues that this web is an *embodied* web: "To be sure, this web is no less bound to the objective world of things than speech is to the existence of a living body, but the relationship is not like that of a facade or, in Marxian terminology, of an essentially superfluous superstructure affixed to the useful structure of the building itself" (183). Although largely unexplored, this passage reveals that Arendt's all-too-tidy distinction between the first birth, the birth of the physical body, and the second, linguistic birth, is eventually undone in her thought. Linguistic natality cannot be laid over physical natality, and this suggests that both births are inseparable and always found together.

Arendt makes the same point early on in *The Human Condition* in her discussion of the meaning of mortality: "Men are 'the mortals,' the only mortal things in existence, because unlike animals they do not exist only as members of the species whose immortal life is guaranteed through procreation. The mortality of men lies in the fact that individual life, with a recognizable life-story from birth to death, rises out of biological life" (19). The difference between the immortal life of the species and the mortality of human beings, Arendt argues, lies in the difference between cyclical and rectilinear temporality. The latter, she argues, "cuts through the circular movement of biological life. This is mortality: to move along a rectilinear line in a universe where everything, if it moves at all, moves in a cyclical order" (19). Mortality is the cut in the temporality of biological life, and yet mortality is not completely cut off from biological life. Instead, the cut of mortality introduces another temporality into biological life, thereby infusing biological life with mortality. Insofar as the event of natality is for Arendt also the event of mortality, it must be understood as always already introducing this mortal cut into biological life, thereby confounding any strict distinction between the biological and the mortal. In response to Flynn, one must say that the event of natality transforms the natural into the mortal. This means that our bodily existence is never simply biological but always already mortal. Here Arendt is very close to Hans Jonas's description of embodiment as already infused with the rectilinear time of the mortal individual: "My identity is the identity of the whole organism. . . . How else could a man love a woman and not merely her brains? How else could we lose ourselves in the aspect of a face? Be touched by the delicacy of a frame? It is this person's, and no one else's."[26] While we have yet to ex-

amine fully Arendt's principle of humanity that emerges in the event of natality, we can see at least one of the risks that (according to Ignatieff) is
courted when grounding human rights in humanity; namely, that "it lacks
any ability to limit the human use of human life" in such instances as biological engineering.[27] While this complex issue is beyond the purview of
the discussion here, the mortality of biological life would provide the limit
for such technological experimentation.

Before turning to linguistic natality, it is important to note that Arendt's
notion of public spaces (which must always be said in the plural)[28] is,
therefore, larger than simply a set of laws and institutions, the latter constituting the political as opposed to the public. As spaces of action, public
spaces are spaces for the possibility of enactive inaugurative speech. In
other words, the polis understood as a space of appearance is performative,
wherein what appears is the unexpected and unpredictable words of a self
who is always in a web of relationships. Only in this way, Arendt argues,
can we understand the meaning of Pericles' words: "Wherever you go, you
will be a *polis*":

> [These words] expressed the conviction that action and speech create a space
> between the participants which can find its proper location almost any time
> and anywhere. It is the space of appearance in the widest sense of the word,
> namely, the space where I appear to others as others appear to me, where
> men exist not merely like other living or inanimate things but make their
> appearance explicitly. (198–199)

The polis as a space of action appears primarily through the performance of the word. Often overlooked in Arendt's thought is the primacy of
speech in the constituting of political action. In *The Promise of Politics*,
Arendt is clear that for her speech is a form of action. Indeed, for her there
is no separation between speech and action. Words belong to the event itself. Looking to the Greek *polis*, she claims that "from the very beginning,
which means as early as Homer, such a separation on principle between
speech and action does not occur—and not only because great words were
needed to accompany and explain great deeds that would otherwise fall
into mute oblivion, but also because speech itself was from the start considered a form of action" (*PP*, 125).

Arendt states in *The Human Condition* that "the sign of the political is,
moreover, not invested in the character of the story itself, but only in the
mode in which it came into existence" (186). In other words, a political
space is constituted in the realm of representation and the process of signification. Arendt therefore has a considerably larger notion of the political
than she is usually given credit for having. While much attention is paid to
Arendt's problematic distinction between the social and the political, the

inseparability of culture and politics in her work is largely overlooked. In her essay "Crisis in Culture," Arendt writes: "Culture and politics, then, belong together because it is not knowledge or truth which is at stake, but rather judgment and decision, the judicious exchange of opinion of public life and the common world, and the decision what manner of action is to be taken in it, as well as to how it is to look henceforth, what kind of things are to appear in it" (*BPF,* 223). The inseparability of culture and politics is indicative of this larger sense of the political as the realm of representation and the process of signification wherein identities of the "who" (and these are always both unique and plural) are reinscribed and re-presented within the web of relationships that constitute the *inter-esse* of public spaces.[29]

A public space, understood as a space of inaugurative speech, allows us to better understand Arendt's claim that the use of clichés and the rise of ideology are foreclosures of such spaces. Clichés disclose nothing and no one; they are clichés precisely because they say nothing. Indeed, it is worse—saying nothing, the cliché darkens a public space by making it more difficult, if not impossible, for the new and unpredictable to appear. Her reference to Heidegger's insight that "the light of the public obscures everything" makes this clear: what ought to be illuminating—namely, the appearance of speech—darkens and obscures whenever speech becomes a cliché. The rise of ideology forecloses a space of inaugurative speech precisely because it systematically sets out to destroy any possibility of the unpredicable or the new by insisting on "strident ideology." This stridency is, for Arendt, tied to a notion of truth as productive and technological:

> If Western philosophy has maintained that reality is truth—for this is of course the ontological basis of *adequatio rei et intellectus*—then totalitarianism has concluded from this that we can fabricate truth insofar as we can fabricate reality; that we do not have to wait until reality unveils itself and shows us its true face, but can bring into being a reality whose structures will be known from the beginning because the whole thing is our product. (*EU,* 354)

"Strident logicality" appeals to the inherent worldlessness and lack of speech that characterizes the modern world of technology. Rather than action, there is only atomization and a "perfect functionality." At the extreme, perfect functionality becomes terror—the elimination of the new. In its place is the "tranquility of the cemetery" (*OT,* 348).

The possibility of the unexpected word therefore reveals the vulnerability and infelicity at the very heart of language. But how can we think this infelicity, this vulnerability that allows for the unexpected word, the new beginning? Or, more precisely, how can we think this vulnerability of language that is the vulnerability of the newcomer, the stranger? And further, what is the relation between the second birth—that is, the unexpected word—and the first birth, that which Arendt calls "the naked fact of physical appearance"? Certainly Arendt claims inseparability between the first and second births, pointing out that it is because of the *fact* of the beginning that the self is able to be born again as the unexpected word. Here Arendt's references to Heidegger are helpful in illuminating the ontological status of the event of natality. Two references especially suggest that Heidegger's thought provides at the very least an implicit frame for Arendt's thought of natality: a reference to his notion of the *Augenblick* and a reference to his statement that "man can speak only insofar as he is a sayer." Both reveal the way language and temporality are at work in the event of natality.[30] These references suggest that despite her claim in *The Human Condition,* the event of natality for Arendt is never simply "the naked fact of physical appearance."

Most significant, Heidegger's discussion of the *Augenblick* first occurs in *Being and Time* in a discussion of the birth and death of Dasein: "Factical Dasein exists as born; and as born, it is already dying, in the sense of Being-towards-death. As long as Dasein factically exists, both the 'ends' and their 'between' *are,* and they *are* in the only way which is possible on the basis of Dasein's Being as *care.*"[31] Dasein's natality is inseparable from its fatality and both are part of Dasein's structure as care (*Sorge*): "As care, Dasein is the 'between'; it is between its no-longer and not-yet." And yet the no-longer and the not-yet are not to be understood as external; Dasein's care functions as bookends. The "between" is permeated by the ends: "Understood existentially, birth is not and never is something past in the sense of something no longer present-at-hand; and death is just as far from having the kind of Being of something still outstanding, not yet present-at-hand but coming along." Dasein's "who," its self-constancy, can therefore be grasped only by understanding a "specific temporalizing of temporality."[32] This leads Heidegger to an analysis of Dasein's authentic historicity, which, he shows, is the temporality of the *Augenblick.* In what follows, I will examine two important aspects of this discussion that illuminate the beginning inherent in our being-born and allow for the possibility of another beginning, or what Arendt calls a "second birth." The first aspect of the discussion is our being-born as *Sorge,* and the second is the temporality of natality understood as the temporality of the *Augenblick.* (This last will also add to our understanding of the temporality of natality.)

Significantly, in *Being and Time,* the first discussion of *Sorge* is given in the discussion of *Fürsorge,* a solicitude that frees. There is an inseparability between natality, freedom, and solicitude. *Fürsorge* does not leap in and dominate the other; instead, it is the solicitude that lets the other be "who" he or she is. What interests us here is how natality, inseparable from fatality, allows for the possibility of our being as care. In other words, the event of natality allows for the freeing solicitude that marks our very being. This contributes to Arendt's claim that the event of natality carries within it the principle of freedom, the possibility of beginning something new.

Heidegger is careful to show in section 41 of *Being and Time* that this act of freedom is not located in the individual will: "Care is always concern and solicitude, even if only privately. In willing, an entity which is understood—that is, one which has been projected upon its possibility—gets seized upon, either as something with which one may concern oneself or as something which is to be brought into it Being through solicitude (*Fürsorge*). . . . In the phenomenon of willing, the underlying totality of care shows through."[33] (In this discussion of authentic care and solicitude, a solicitude that frees rather than dominates, Heidegger first formulates what he will much later think as *Gelassenheit*, "the will not to will.") The activity of freedom is the activity of receptivity. It is not an act of the will but of the word—the finite word that receives—that welcomes—the newcomer and the stranger. It is the initial power of the welcoming address that inaugurates and sustains linguistic existence, conferring singularity in time and place. The facticity of the self is given through the welcoming word. In other words, in the event of natality, our thrownness (*Geworfenheit*) is always bound up with a welcoming. The self, set free through the address of the other, becomes a self capable of addressing others.

One comes to be through the welcoming word. To be named is to be welcomed—and this is the entrance into language. This is precisely what Arendt suggests when she refers to Heidegger's statement that "man can speak only insofar as he is the sayer."[34] The possibility of saying is the welcoming. Furthermore, Heidegger suggests that this entrance into language is an *embodied* entrance. Indeed, Heidegger's analysis sheds light on Arendt's undeveloped claim in *The Human Condition* that speech is not merely an essentially superfluous superstructure affixed to embodiment. In this section of *Being and Time*, Heidegger suggests that the first birth, the being-born is never simply the naked fact of physical existence but is already infused with language. In fact, *Fürsorge* carries with it the sense of "prenatal care," rendering impossible a distinction between the word of welcome and the fact of physical birth.[35] In Arendtian terms, *zoe* is always already *zoon logon ekhon*. Naked facticity is always already the site of language— that is, the address that calls one into existence. Article Seven of the International Convention of the Rights of the Child explicitly articulates this insight: "The child shall have the right from birth to a name."

Moreover, Heidegger's discussion of *Entfernung*—that is, Dasein's embodiment as always at a distance from itself—suggests that embodiment, "naked facticity" is never fully present; it is itself permeated by a lack and loss that marks the event of natality. This is extremely important insofar as it would seem that one reason Arendt separates *zoe* from the *bios politikos* is in order to ensure that the political space is not understood in terms of some organic or naturalistic fantasy that allows for the totalitarian attempt to make this fantasy present. However, in his discussion of *Raumlichkeit* (better translated as embodiment), Heidegger offers a way to think the materiality of the political body that disallows any kind of organic fantasy. *Raumlichkeit* denotes our embodied spatiality. We are, therefore, embodied beings, always already immersed and involved in the world. The question of the place of our embodiment, however, is not a question of a corporeal thing present-at-hand. Being-alongside is an embodied "here" that is also "there": "Dasein understands its 'here' in terms of its environmental 'yonder'. The 'here' does not mean the 'where' of something present-at-hand but rather the 'whereat' (*Wobei*) of deseverant Being-alongside together with this deseverance."[36]

Deseverance (*Entfernung*) means that one is always cut off from fully possessing one's embodied being. Moreover, Dasein's embodiment is never the where in which Dasein stands. The place of embodiment is the *Wobei;* it is a region given in the body's directionality: "To free a totality of involvements is equiprimordially, to let something be involved in a region, and to do so by deseverance and directionality."[37] Directionality allows for orientation; for example, the difference between left and right. The difference, Heidegger maintains, against Kant, is not given in a subjective feeling; rather, orientation is always already given in an involvement whose basis is the deseverance and directionality of embodiment. The point here is that Dasein's embodied stance (*Wobei*) is always given within a region of involvements, wherein Dasein is always immersed and always finds itself from a distance. Most important, however, the there is never crossed out completely and therefore there is never any possibility of an embodied presence-at-hand. Dasein's embodiment is the place of difference and dislocation. Significantly, Heidegger offers a way to think *zoe* and *bios politikos* as inseparable while at the same time avoiding any notion of an organic, totalizing fantasy.

Heidegger shows that to the being-born and the dying that are without place or representation, this address, this hearkening, is always aimed at that which is without a why; it is to be addressed or called at the moment when we do not or no longer have possessions or identity cards to present. It is a saying that says nothing other than the lack or loss out of which its saying is articulated. This is *Verschwiegenheit,* the silence that articulates Dasein's potentiality-for-being,[38] the articulated silence that is the voice of the friend, the welcoming that every Dasein carries with it. There is nothing so "other" than being born and dying, the indices of all alterity. There is nothing that makes clearer gratitude for being received: *Fürsorge.* And is this not the gratitude that marks Arendt's own thinking of gratitude inherent in the event of natality?

The event of natality therefore marks a limit in the sense of an opening of our possibilities. The opening is never at our disposal. We cannot set our own limits and cannot absorb the beginning and the end into those limits.[39] In short, through the event of natality, we are held in an opening we did not create and can never appropriate. Indeed, as the stretch between natality and fatality, a stretch that is itself permeated by these nonappropriable ends, existence is permeated with loss, a radical alterity to which there is no relation. The originary event proceeds from nothingness; it is an absolute beginning, just as our death is an absolute end entirely without relation. The political significance of this is that the political space, the space of beginning something new, is always open to something other than itself.

This last allows us to see how the temporality of the first birth is always already at work in the second birth. Arendt explicitly refers to both Heidegger's notion of the *Augenblick* and Nietzsche's eternal return in her discussion of Kafka's parable as she attempts to articulate the meaning of the temporal duration of the gap wherein the "who" stands, arguing that this is the moment of the untimely.[40] Arendt, however, does not explicitly link Heidegger's discussion of the *Augenblick* to her concept of natality, which is characterized by the temporality of this gap. If she had done so, she would have been faced with the way in which our naked facticity is always already imbued with linguistic natality.[41] Indeed, this is precisely what Heidegger shows in his discussion of the temporality of the *Augenblick* in *Being and Time.* He prefaces the discussion by stating that freedom and finitude are at the very heart of this temporality of the *Augenblick:*

> Only an entity which, in its Being, is essentially **futural** so that it is free for its death and lets itself be thrown back upon its factical "there" by shattering itself against death—that is to say, only an entity which, as futural, is equiprimordially in the process of **having-been,** can, by handing down to itself the possibility it has inherited, take over its own thrownness and be **in the moment of vision** for "its time." Only authentic temporality, which is at the same time finite, makes possible something like fate—that is to say, authentic historicality.[42]

Within the moment of insight (*Augenblick*), one grasps historical, factical possibilities. At the same time, the moment of insight is a site, a historical conjunction of the temporal and spatial. One's resolute, authentic response to one's historicity is possible only at this moment of insight in the taking over of *repeatable* historical possibilities.

It is the notion of repetition that I now wish to explore here—what kind of repetition is at work in the *Augenblick*? Heidegger insists that the repetition cannot be understood as mere repetition, a simple taking over of historical possibilities. The word Heidegger uses is "*erwidert.*" This word must be understood from its preposition "*wider,*" meaning "contrary to or against":

> The repeating of that which is possible does not bring again (*Wiederbringen*) something that is "'past,'" nor does it bring the "Present" back to that which has already been "outstripped." . . . Rather, the repetition makes a *critical reply* (e*rwidert*) to the possibility of that existence which has-been-there. But when such a rejoinder is made to this possibility in a resolution, it is made in a *moment of vision*; *and as such* it is at the same time a *disavowal* of that which in the "today" is working itself out as the "past."[43]

The "between" marking the temporality of the *Augenblick* is the time of this repetition that takes the form of a reply that is at odds with what has been handed down. Indeed, repetition understood as "critical response" (*Erwiderung*) is what allows for the unexpected and unpredictable, precisely because it disturbs and disrupts. We need to recall here Arendt's reading of Benjamin, especially the Benjaminian displacement or diffraction of the present. The critical response dissolves any authorization of repeatable possibilities based on a myth of being tied to a beginning; it is possible because the "between" that marks the temporality of the *Augenblick* (and let us not forget that this is the "between" that marks the "who" of Dasein) is a gap between the no-longer and the not-yet, a gap that is permeated by this loss or lack that marks both ends. Thus, the moment is untimely and out of order. Paradoxically, the temporality of the *Augenblick* reveals that the constancy of the self is always becoming *undone* in the moment of the critical reply. As the "between," the "who" is untimely, out of order. And this is because the lack, the loss that marks being-born and dying, permeates every moment of the being.

The significance of this loss at the very heart of speech for our freedom and finitude is now clear. The first beginning, being-born, makes possible a second beginning, if by that is understood the possibility of saying something new. The latter is the Arendtian "second birth." At the beginning of saying, there is a loss, a lack at the very heart of the reply, which means that the possibility of a different reply is never foreclosed. While speech is always historically situated and delimited, it is also open to further unexpected delimitations. The temporality of the *Augenblick* is the moment in which a speech without prior authorization nevertheless can assume authorization in the course of its saying. The loss or lack that permeates the reply reveals that the word always has the possibility of being unmoored. The force of the critical reply is derived precisely from its decontextualization, from its break with a prior context and its capacity to assume new contexts. Limited by natality and fatality, speech is bound to nothing. Although materially entangled, it is bound to no context in particular. Thus, there is an infelicity, a vulnerability to all enacted speech—an inherent independence from any of the webs in which it appears. There is always the possibility of the unexpected word.

Although largely ignored by readers of both Heidegger and Arendt, Heidegger's notion of being-toward-birth, particularly when thought through his discussion of *Fürsorge,* offers a correction to Arendt's discussion of natality, a correction not entirely unsanctioned by Arendt herself insofar as her references to Heidegger's notion of the *Augenblick* as well as his understanding of language point directly to his discussion of natality. Heidegger shows how the "naked fact of physical birth" is always already linguistically bound. To be born, to be a mortal, is to have been welcomed, to have been given a name. And only because the "who" is named is he or she able to die. It is the welcoming through the name that renders one mortal. At the same time, Heidegger's account of natality complicates Arendt's strict separation between *zoe* and *bios politikos* with her subsequent distinction between the social and the political. Heidegger's account of *Fürsorge* and being-toward-birth articulates how it is that the "second birth" is never simply laid over the first. Instead, both births happen at once.

THE DOUBLE *PRINCIPIUM* OF NATALITY

Rejecting "substance ontology," Arendt develops instead an "ontology of the event." Answering her own question as to whether the unprecedented condition of global unity points to a ground, until now hidden, of the human condition on earth, she points to the event of natality as the fundamental event of the human condition and the source of the principle of the 'right to have rights.' For Arendt, the *principium* of the archaic event of natality is double: the principle of *initium* and the principle of givenness. This double principle has tremendous consequences for our understanding of the ontological foundation of human rights. Recalling our discussion of Montesquieu, the double principle will orient human rights in ways that at first seem incompatible with Arendt's more explicit discussion of the public space.

Arendt's debt to Augustine's insight that the event of natality carries the promise of beginning is well known. The final optimistic words of the otherwise deeply pessimistic *The Origins of Totalitarianism* are those of Augustine's: *Initium esse homo creatus est*—that a beginning was made man was created. "This beginning is guaranteed by each new birth; it is indeed every man" (*OT,* 479). Natality understood in terms of *initium* provides the *arche* for Arendt's subsequent work, informing her key political concepts of action, freedom, and power. Indeed, our capacity to appear and to begin is at the root of our pleasure in excelling in the company of others.

There is, however, an earlier reference in *The Origins of Totalitarianism* to Augustine's understanding of natality that Arendt never fully develops but that points to another dimension of the event of natality and to a second principle, inseparable from the principle of *initium,* also at work in the right to have rights. This earlier reference occurs in Part Two of *The Origins of Totalitarianism* at the conclusion of her analysis of imperialism, which ends with an examination of the decline of the nation-state and human rights. In the very last pages of this analysis, Arendt describes those stateless refugees (including herself) who, having lost their political status as citizens, have also lost any and all recourse to human rights. It is in the context of the loss of human rights that Arendt refers to Augustine. I quote the text at length:

> The human being who has lost his place in a community, his political status in the struggle of his time, and the legal personality which makes his actions and part of his destiny a consistent whole, is left with those qualities which usually can become articulate only in the sphere of private life and must remain unqualified, mere existence in all matters of public concern. This mere existence, that is, all that which is mysteriously given us by birth and which include the shape of our bodies and the talents of our minds, can be adequately dealt with only by the unpredictable hazards of friendship and sympathy, or by the great and incalculable grace of love, which says with Augustine, "*Volo ut sis* (I want you to be)" without being able to give any particular reason for such supreme and unsurpassable affirmation. (*OT,* 301)

Here Arendt points to another dimension of the event of natality, pointing approvingly to the Augustinian insight that the event of natality is *also* about that which is given. As I will argue more fully in the next two chapters, these two principles, the principle of *initium* and the principle of givenness, provide the principle of humanity that founds the right to have rights.

TWO

The Principle of *Initium:*
Freedom, Power, and the
Right to Have Rights

> The beginning is like a god which as long as it
> dwells among men saves all things.
> Hannah Arendt, *Between Past and Future*

In the preface to *The Origins of Totalitarianism,* Arendt argues that the task of thinking today lies in "bearing consciously the burden which our century has placed on us—neither denying its existence nor submitting meekly to its weight." Comprehending, she claims, means the "attentive facing up to, and resisting of, reality—whatever it may be" (*OT,* viii). Nowhere is this attempt at comprehension more evident than in Arendt's facing up to her own experience as a refugee who, having lost her status as a citizen, lost all claim to human rights. At the very moment when protection under the auspices of universal human rights was most desperately needed, no such protection was granted. Outside the law and not belonging to any political community, she and her fellow refugees were reduced to "mere naked human beings" in a "condition of complete rightlessness" (296). Most devastating, she argues, is that the world found nothing sacred in the "abstract nakedness" of being human: "If a human being loses his political status, he should, according to the implications of the inborn and inalienable rights of man, come under exactly the situation for which the declaration of such general rights provided. Actually the opposite is the case. It seems that a man who is nothing but a man has lost the very qualities which make it possible for other people to treat him as a fellow-man" (300). Indeed, she points out, the gas chambers were set in motion by first depriving Jews "of all legal status (the status of second-class citizenship) and cutting them off from the world of the living by herding them into ghettos and concentration camps. . . . The point is that a condition of complete rightlessness was created before the right to live was challenged" (296).

Facing up to the condition of being reduced to a mere human being, a situation in which she literally had to flee for her life, Arendt offers a radical critique of the modern formulation of human rights. She argues that these supposedly inalienable universal human rights were from their inception inseparably tied to the sovereignty of a people. Her critique and reformulation of the modern understanding of human rights in large part rests on a critique and reformulation of the notions of freedom and agency at the very heart of modern human rights discourse, which, she argues, are framed in terms of sovereignty, both individual and collective. Indeed, Arendt's location of freedom and justice in the more foundational issue of the right to have rights goes far in answering her critics, such as Dana Villa, who charge Arendt with ignoring the liberal tradition, especially its notion of justice. In her discussion of modern human rights, Arendt does not dismiss the liberal tradition; rather, she shows how this tradition, with its paramount concern for freedom and justice, does not grasp that politically there is something more fundamental: "Something much more fundamental than freedom or justice, which are the rights of citizens, is at stake when belonging to the community into which one is born is no longer a matter of course and not belonging no longer a matter of choice, or when one is placed in a situation where, unless he commits a crime, his treatment by others does not depend on what he does or does not do. This extremity, and nothing else, is the situation of people deprived of human rights" (296). For Arendt, more fundamental than the rights of justice and freedom is the right to action and opinion and the right to belong to a political community in which one's speech and action are rendered significant: "We became aware of the existence of a right to have rights (and this means to live in a framework where one is judged by one's actions and opinions), as well as the right to belong to some kind of organized community, only when millions of people emerged who had lost and could not regain these rights because of the new global political situation" (296–297).

Arendt's theoretical reformulation of the fundamental right to have rights emerges out of her reflection on the *initium* inherent in the ontological event of natality that makes every human being a beginner. Arendt is indebted to Augustine's insight "*Initium ergo ut esset, creatus est homo, ante quem nullus fuit*" (That there might be a beginning, man was created before whom nobody was) (*EU,* 321). This principle of *initium,* she argues, allows for a radical reformulation of the modern framework of human rights such that the rights of freedom and agency are rooted in the more fundamental right of action and speech. Moreover, the right of sovereignty, individual and collective, is replaced with the right to belong to an organized political space, with its inherent plurality of actors.

SOVEREIGN AGENCY AND THE NATION STATE:
THE MODERN CONCEPTION OF HUMAN RIGHTS

The location of human rights in a metaphysical notion of an autonomous sovereign subject plays the leading role in the modern conception of human rights. In this framework, human rights are viewed as inalienably possessed by sovereign subjects understood as bearers of rights. The two hallmarks of modern human rights, therefore, are individual agency and self-determination. Although Arendt's readers have focused almost exclusively on her analysis of Edmund Burke's conception of "citizen's rights," Hobbes and Rousseau are for her the tradition's principle theorists. While Hobbes denies the presence of free will in a world where all movement is causally determined, he nonetheless defines *jus naturale*, natural right, as the "Liberty each man hath, to use his individual power as he will himself, for the preservation of his own Nature, that is to say, of his own Life; and consequently, of doing any thing, which in his own Judgement and Reason, he shall conceive to be the aptest means thereunto."[1] Hobbes locates natural right in the individual's own power and movement to do what he can do to preserve his own life. Hobbes defines liberty as the "absence of external Impediments, which Impediments, may oft take away part of a man's power to do what he would, but cannot hinder him from using the power left him, according to his judgment, and reason shall dictate to him."[2] However nasty, brutish, and short an individual's life might be, in the state of nature each is sovereign, with the natural right to everything, including another's body. Hobbes is clear that conditionally laying down one's right to everything in the interest of seeking peace and security does not result in another's rightful gain: "For he that renounceth, or passeth away his Right, giveth not to any other man a Right which he had not before; because there is nothing to which every man had not Right by nature: but only standeth out of his way, that he may enjoy his own originall Right, without hindrance from him; not without hindrance from another."[3]

For Hobbes, liberty and power are synonymous terms: the natural right to self-preservation is understood as a kind of "natural power." Natural right is the inalienable power or liberty each individual possesses to move without hindrance in order to preserve his or her life. As is well known, Hobbes defines power in *Leviathan* as the "present means to secure the future." Power is one's present means to secure one's future self-preservation, which means one's future security. As Arendt points out, this is power for the sake of power (*OT,* 141). Everything—whether in the form of knowledge or wealth—is reduced to power: "Therefore, if man is actually driven by nothing but his individual interests, desire for power must be the fundamental passion of man" (139). In the Hobbesian framework, she argues, natural right corresponds to the endless process of accumulating power. And the equality of human beings lies solely in their equality of cunning in obtaining more power. This equality therefore has no inherent worth. As is well known, for Hobbes it is not the seller but the buyer who determines the worth of an individual, worth being determined by the amount of power one has at one's disposal. Arendt points out that for Hobbes the individual possesses no inherent dignity worthy of respect; instead, worth is dependent upon power, which is determined solely in the eyes of others.

Hobbes therefore excludes in principle the idea of humanity, an exclusion that Arendt argues had disastrous effects in the nineteenth century, when his philosophy provided the underpinnings of race ideology: "The philosophy of Hobbes, it is true, contains nothing of modern race doctrines, which not only stir up the mob, but in their totalitarian form outline very clearly the forms of organization through which humanity could carry the process of capital and power accumulation through to its logical end in self-destruction. But Hobbes at least provided political thought with the requisite for all race doctrines, that is, the exclusion in principle of the idea of humanity which constitutes the sole regulating idea of international law" (157). For Arendt, the idea of humanity is the sole constituting basis of human rights, which, in turn, provides the basis for international law. Here it is important to note that for Arendt the exclusion of the idea of humanity and the reduction of human rights to the self-interested power of a sovereign and isolated individual provides the theoretical underpinnings first to nineteenth-century imperialism, in which "everything is permitted," and then to twentieth-century totalitarianism, with its race ideologies, for which "everything is possible."

Significant for Arendt's reformulation is that fact that in the Hobbesian framework of natural right, reason is private; it is the reason and judgment of an individual who calculates the best way to obtain its *own* self-interest, the most fundamental being the interest in self-preservation. Reason, the "spy and scout for the passions," is inseparable from individual desire for self-preservation. Indeed, Hobbes goes so far in *Leviathan* as to argue that the fundamental desire for self-preservation acts as a lawful imperative for action: "The language of Desire, and Aversion, is Imperative; as Do this, forebear that, which when the party is obliged to do, or forbear is Command, otherwise Prayer, or else Counsel."[4] Distinguishing between the imperative and optative modes (the subjunctive mode expressing wish [*optare*]: "Had I the means, this would be done"), Hobbes argues that the passions are imperative.[5] They command: do this, forbear that. Thus, there is obligation or lawfulness at the level of the two fundamental passions: desire and aversion. Insofar as these passions are always mediated by signification, they are always already linked to reason and the general imperative: seek peace and follow it.

This is the significance of Hobbes's distinction between *in foro interno* and *in foro externo*. Hobbes argues that the two fundamental passions, appetite and aversion, command *in foro interno,* although not always *in foro externo.* That is, they oblige merely to the extent of the desire that they should take place but not necessarily to extent of the act.[6] The rational laws of nature, which arise from and are obligated to the imperative of desire, oblige us only *in foro interno.* Arising out of the command of desire, the fundamental law of nature is a general directive or principal to seek peace. Through the imperative of desire, one is obligated rationally to seek peace. The imperative of desire commands that one bring about a situation in which the condition for peace is fulfilled—that is, the renunciation of the right to arrogate everything to oneself. *In foro interno,* these laws establish no specific obligations but instead direct us to bring about a situation in which specific obligations will arise. (Of course, the ultimate situation will be the establishing of the absolute sovereignty of the commonwealth.) Important here is reason's obligation to the imperative of desire. As the scout and spy of the passions, the laws of reason are inseparable from the commands of desire. Further, the rational obligation to renounce the right to everything and to seek peace arises out of the command of the desire for self-preservation. The natural laws of reason are from the beginning tied to a self-interested ego that calculates the best way to meet this obligation of desire.

Hobbes therefore articulates natural right in terms of the self-interested individual whose fundamental interest is the power to obtain future security. For Hobbes, rights are private rights that correspond to individual freedom of action located in an autonomous sovereign subject. Reason is the private reason of sovereign individuals whose sole purpose is the reckoning of their own interests. As Arendt points out, "Thus membership in any form of community is for Hobbes a temporary and limited affair which essentially does not change the solitary and private character of the individual (who has 'no pleasure, but on the contrary a great deal of griefe in keeping company where there is no power to overawe them all') or create permanent bonds between him and his fellow-man" (140). For Arendt, the sovereign commonwealth follows from Hobbes's reduction of natural right to the power of a sovereign self-interested individual who initially has the right to everything. The sovereign power of the Commonwealth, she argues, is made up of private individuals solely interested in the desire for power; it embodies the sum total of private interests: "Hobbes' *Leviathan* exposed the only political theory according to which the state is based not on some kind of constituting law—whether divine law, the law of nature, or the law of social contract—which determines the rights and wrongs of the individual's interest with respect to public affairs, but on the individual interests themselves, so that the 'private interest is the same with the publique'" (139). Delegation of power (synonymous with right) flows from the absolute power of the state as monopolized power that demands absolute obedience:

> The Commonwealth is based on the delegation of power, and not of rights. It acquires a monopoly in killing and provides in exchange a conditional guarantee against being killed. Security is provided by the law, which is a direct emanation from the power monopoly of the state. . . . And as this law flows directly from absolute power, it represents absolute necessity in the eyes of the individual who lives under it. (139)

The reduction of right to the sovereign power of the state follows directly from the sovereign power of self-interested individuals. The sovereign himself ensures an "ordered system of egoisms" of individuals who have agreed to give up individual sovereignty in exchange for a sovereign that can better secure the self-interests of each.[7] Arendt's critique of the modern conception of human rights is to a large extent carried out through a critique of sovereignty first developed by Hobbes, which she argues is at the very heart of the modern formulation of right.

In her analysis of Rousseau, Arendt continues to develop her critique of the modern formulation of human rights in which sovereign power and right are inexorably linked. She argues that while it might seem as if Rousseau rejects Hobbes's reduction of right to sovereign state power, arguing instead for a concept of natural right located in subjective rights grounded in subjective freedom, this is not the case. While Rousseau argues vehemently against Hobbes's reduction of right to power, Arendt claims that by locating freedom first in the subjective will of an individual agent and then in the general will of a people, Rousseau continues the Hobbesian tradition of ultimately reducing human rights to sovereign state power. Her critique of Rousseau is instructive not only for grasping her critique of the modern conception of human rights but also for understanding how her conceptions of freedom, power, and acting allow her to offer a new theoretical framework for human rights. Only through a radical reformulation of power, freedom, and the public space, Arendt argues, is it possible to sever human rights from sovereign agency and sovereign state power.

Arendt's analysis of Rousseau focuses on his location of natural rights in the free will of an autonomous and sovereign subject. As is well known, arguing against the natural sociability of human beings, Rousseau establishes natural right on the basis of the two principles or sentiments written on the human heart; namely, the sentiment of one's own existence and well-being and the sentiment of pity—that is, our natural repugnance at the suffering of other sentient beings.[8] Insofar as self-preservation is something human beings share with other animals, Rousseau argues, contrary to Hobbes, that the sense of "well-being" is the natural sentiment of our freedom, which, he argues, is a higher good than life itself.[9] Arendt agrees with Leo Strauss's assertion that Rousseau can be called the "first philosopher of freedom" insofar as for Rousseau the individual free will is a good in itself: freedom is the fundamental good and is that which defines the nature of the human being in its independence from other human beings.[10] Against Grotius's claim that rights are alienable and can be bought and sold (and have in fact been forfeited by many peoples, each of which has chosen temporary security by capitulating to an arbitrary governmental authority), Rousseau claims that rights are inalienable because freedom is what distinguishes humans from other animals. Hence it follows that freedom is a natural inalienable right located in the individual will. In the *Discourse on the Origin of Inequality,* Rousseau argues that political rights follow from this natural right to freedom and self-determination: "It is therefore incontestable and it is the fundamental maxim of all political right, that peoples have given themselves chiefs to defend that freedom and not to enslave themselves."[11]

At the same time, Rousseau argues that a second principle, pity, is at work in grounding human rights. Pity is a sentiment by which others are present in a way that is neither threatening nor malicious; it is a sensibility by which an individual is aware of and touched by the feelings and attitudes of others. From the conjunction of these two principles or sentiments, Rousseau establishes natural right "without it being necessary to introduce into it that of sociality." Anterior to society, the natural human being is endowed with the natural right to preservation of his or her own well-being and freedom and at the same time limited by the natural pity each individual feels at the suffering of others. Natural right includes a kind of negative liberty that limits each individual's preservation of his or her own life and natural independence. Our natural sentiment of our own existence and natural independence is always accompanied by a sentiment of the other's existence and independence. Natural pity does not constitute a natural sociability insofar as the state of nature is a state of natural independence, but it does place a limit on the individual's natural freedom. Rousseau argues that because human beings are limited by natural pity, there is no need to make a contract with one another to restrict our right to everything. Instead, the task of the social contract is to institute politically the natural (and limited) independence of each individual. In *The Social Contract,* Rousseau states, "But the social order is a sacred right, which provides the basis for all the others. Yet this right does not come from nature; it is therefore founded on conventions."[12]

Like Hobbes, Rousseau argues that these two principles establishing natural right are anterior to reason, although it is the role of reason to reestablish politically the principles of natural right. (To be sure, Hobbes and Rousseau differ greatly on the nature of the sentiment, the first arguing for the desire for self-preservation and the second for the twin sentiments of freedom and pity.) In the *Second Discourse,* Rousseau refers to pity as a "virtue all the more universal and useful to man as it precedes the exercise of all reflection in him, and so Natural that even the Beast sometimes shows evident signs of it."[13] He goes on to state: "Such is the pure movement of Nature prior to all reflection; such is the force of natural pity."[14] In *Emile,* Rousseau develops his thinking on the pre-reflective sentiment of pity: "I am, so to speak, in him, it is in order not to suffer that I do not want him to suffer. I am more interested in him for love of myself, and the reason for the precept is in nature itself, which inspires in me the desire of my well-being in whatever place I feel my existence. Precepts of natural law are not founded on reason alone. They have a basis in love of self which is the principle of human justice."[15] Rousseau goes on to argue that "there is in the depths of souls, then, an innate principle of justice and virtue according to which, in spite of our maxims, we judge our actions and those of others as good or bad. It is to this principle that I give the name *conscience.*"[16] Conscience, then, is a natural, pre-reflective voice that Rousseau first articulates as pity. Furthermore, Rousseau claims, "Ideas are acquired, the sentiment of conscience is not."[17] Without the sentiment of conscience, reason itself is incapable of moving human beings to action. Reason, the ability to adopt the general standpoint—to adopt the deliberative standpoint of the general will—has its roots in the sentiment of pity and the natural voice of conscience. Reason, therefore, is practical, articulating the generalized interests of individual self-love (*amour de soi*).

Arendt's criticism of Rousseau focuses on his claim that natural rights are located in the twin principles of freedom and pity, themselves found in an autonomous subjective will. She argues that the location of human rights in a free and sovereign will, first as independent *amour de soi* and then as the uncorrupted *amour propre* that retains its natural independence as a member of the general will of the nation-state, duplicates the Hobbesian framework: human rights are understood in terms of power and sovereignty. Indeed, she argues that this notion of sovereign power is tied to a long tradition that has its origins in Augustine. Of course, for Augustine the problem of being able to act or, more precisely, the inability to act is tied to a will that is divided against itself. This is his dilemma: "I will and I cannot."[18] A powerful will is therefore possible only if the will is one with itself. Arendt points out that the conception of power that is located in a unified will is precisely the conception of power developed in modern political theory, a fact that has serious consequences for the modern formulation of human rights.

Arendt argues that Rousseau's turn to the unanimous will of the nation is motivated by the problem of the profound instability of all modern political bodies, which is the result of an elementary lack of authority.[19] One way to solve the problem of legitimate authority, she argues, is to make the nation absolute. The legitimacy of power and the legality of the laws would reside in the sovereign will of the nation: the general will of the nation that would reflect the innate, natural goodness of each individual heart and will. Because Augustine has already demonstrated that a divided will is impotent, unanimity is mandatory for the concept of a powerful general will. Thus, Arendt contends, the notion of the general will *must* be based on unanimous consent: unanimity of opinion rather than a plurality of opinions. The unanimity of the general will, whose absolute sovereignty guarantees the stability of the political realm, depends upon the individual wills giving up their particular interests and consenting to be ruled by a government whose power has become sovereign *precisely because* individuals have given up their individual power. As in Hobbes, the existence of an absolute sovereign in whom the identical origin of law and power is embodied makes the law powerful and power legitimate. This does not change when the absolute sovereign is the general will of the people.

Thus, the act of consent combines the principle of absolute rulership and national principles "according to which there must be one representative of the nation as a whole, and where the government is understood to incorporate the will of all nationals" (*OR,* 171). Here we see the emerging form of the sovereign nation-state. Indeed, Arendt argues that the modern understanding of the self as a subject who is the bearer of inalienable rights is inseparable from the notion of the sovereign nation-state that gets its power from the sovereign general will of the people. Human rights, then, are tied up with the question of national emancipation: "Only the emancipated sovereignty of the people, of one's own people, seemed to be able to insure them" (*OT,* 291). In this schema, power is always associated with sovereignty and unity, either the unified sovereign will of the individual or the unity of the general will embodied in the figure of the ruler. Napoleon is the exemplary figure: "I am the *pouvoir constituent*" (*OR,* 163).

Arendt argues that the problem with this schema is twofold. First, rights are indistinguishable from the sovereignty of the general will of the nation-state, and this schema does not allow for the rights of those who are not recognized as part of the general will. This leads to the second problem: at the level of the individual, one must be a national in order to have rights. Arendt's critique of the modern understanding of inalienable rights is founded upon this recognition that these rights were from the beginning tied up with national sovereignty. And, she argues, no group saw this more clearly than those who had lost the protection of the sovereign: "The rights of man, supposedly inalienable, proved to be unenforceable—even in countries whose constitutions were based on them—whenever people appeared who were no longer citizens of any sovereign state" (*OT,* 293). When Billy Budd's ship *The Rights of Man* sails off, leaving him an impressed seaman in the British navy, he cries, "Good-bye, Rights of Man!"[20] Arendt echoes that cry. Human beings become singular and concrete by the exercise of human rights tied to state sovereignty, and those who are deprived of them become human beings without portfolio, "human beings in general." The abstract person is precisely one who is deprived of the rights of man. Contrary to Burke, for Arendt, the abstraction is the abstract nudity of those who are nothing other than human beings who are naked as the day they were born. She argues that it was in order to escape this "abstract nudity" that those who were stateless "insisted upon their nationality, this ultimate fact of their being citizens" (300).

Arendt would disagree with Claude Lefort's argument that the space of modern democracy is a space in which the notion of right mutates in such a way that it is no longer tied to a notion of sovereignty. Lefort argues that in modern democracy, human rights are transformed from the "state of right" (*etat de droit*) to the "right of opposition" (*opposition du droit*). The 1798 Declaration of the Rights of Man and the Citizen is a declaration of opposition, of resistance. The state cannot grant this right. Moreover, with this declaration of the right of opposition, Lefort argues, occurs the disincorporation of right and power. The right to oppose power reveals that right is external to power. Insofar as the declaration reveals that the state is no longer a state of right but of opposition, the political space of modern democracy is a theater of contestation. Right becomes the questioning of right. And because right is always in question, the "we" of the people who declare these rights is always in question. Lefort argues that the declaration of rights as *opposition du droit* "invites us to replace the notion of a regime governed by laws, of a legitimate power, by the notion of a regime founded upon the legitimacy of a debate as to what is legitimate and what is illegitimate—a debate which is necessarily without any guarantor and without any end."[21]

Arendt's experience of being a refugee deems Lefort's analysis far too optimistic. While Arendt would agree that the Declaration of the Rights of Man initially opposes itself to the state in the name of the "good people," it never calls into question the sovereignty and legitimacy of "the people." She claims that contrary to the monarchical state, "whose supreme function was the protection of all inhabitants in its territory no matter what their nationality, and [which] was supposed to act as a supreme legal institution," the modern state was reduced to the "will of the people" and was forced to "recognize only nationals as citizens, to grant full civil and political rights only to those who belonged to the national community by right of origin and fact of birth" (*OT,* 230).[22] For Arendt, the declaration establishes the people themselves as the sovereign bearer of right. While it may very well be the case that the people debate their rights among themselves, those who are not considered part of the people have no right to speak. Sovereign power and right remain inseparable. Contrary to Lefort, Arendt would agree with Foucault: modern political theory "has yet to cut off the head of the king."[23]

Here we begin to see the complicated relation Arendt has to Edmund Burke's conception of human rights. Lefort criticizes Arendt for endorsing Burke's position on human rights and thereby confining rights to the gates of the city. Luc Ferry makes the same charge, arguing that Arendt, along with other critics of modernity, is "resolutely unable, precisely because of the critique of subjectivity . . . to think of something like human rights— an inability that writers like Strauss or even Arendt make no effort to conceal; witness, for example, the chapters on human rights in Arendt's *The Origins of Totalitarianism* where she harks back, certainly with nostalgia, but nonetheless with resolution, to the thinking of Edmund Burke."[24] Arendt's position vis-à-vis Burke is far more complicated. It is true that Arendt agrees with Burke's critique of the notion of natural rights. She rejects the French Declaration of the Rights of Man in favor of the U.S. Bill of Rights, arguing that "the first were meant to spell out primary positive rights, inherent in man's nature, as distinguished from the political status, and as such they tried indeed to reduce politics to nature. The Bills [*sic*] of Rights, on the contrary, [was] meant to institute permanent restraining controls upon all political power, and hence presupposed the existence of a body politic and the functioning of political power" (*OR,* 108–109). To reject the notion of natural rights, however, does not necessarily entail the further claim that human rights belong only to citizens of particular nation-states. We saw above that Arendt's criticism of the modern formulation of human rights is precisely that they are inseparable from the condition of citizenship, which from the start denies the universal aspect of these rights. The paradox for Arendt is that while human rights need always to be politically instituted (hence giving her argument a certain proximity to Burke's), nevertheless the political institution of human rights must be established on the basis of a universal principle of humanity that animates the institution. (Arendt's debt to Montesquieu is evident here.) This places her at a far distance from Burke's insistence that the universal human being is an abstract being and hence that there are only the rights of citizens.

Indeed, Arendt places Burke in the group of thinkers who are the predecessors of modern state-based racism. Burke's insistence on reducing rights to the rights of Englishmen, she argues, is dangerously tied to his notion of a "race of aristocratic blue bloods." This is possible, she argues, precisely because of his complete dismissal of a principle of humanity at work in human rights. As we saw above in her critique of Hobbes, racial thinking depends upon this dismissal:

> But Hobbes at least provided political thought with the prerequisite for all race doctrines, that is, the exclusion in principle of the idea of humanity which constitutes the sole basis of international law. . . . If the idea of humanity, of which the most conclusive symbol is the common origin of the human species, is no longer valid, then nothing is more plausible than a theory according to which brown, yellow, or black races are descended from some other species of apes than the white race, and that all together are predestined by nature to war against each other until they have disappeared from the face of the earth. (*OT,* 157)

In her analysis of race thinking before racism, which includes a long section on Burke, she points out that at the heart of race thinking is the consistent denial of the "great principle upon which national organizations of peoples are built, the principle of equality and solidarity of all peoples guaranteed by the idea of mankind" (161). Race, she argues, marks not the origin of humanity but its end: "Race, politically speaking, is not the beginning of humanity but its end, not the origin of peoples but their decay, not the natural birth of man but his unnatural death" (157). As we have seen, for Arendt, the common origin of humanity lies not in any naturalistic beginnings but in the archaic beginning that marks the event of natality. Paradoxically, this original event that provides the universal principle of humanity is at the same time the origin of unpredictable singularity. In other words, this event carries with it the principle of solidarity even as it gives birth to the singular and the unique. For Arendt, the universal principle of humanity is not abstract precisely because it is given in the archaic event of natality, whereby the unique and singular human being makes its appearance in the world.

FREEDOM, AGENCY, AND RIGHTS:
IGNATIEFF AND HABERMAS

Before turning to Arendt's formulation of the principle of humanity, we must consider first just how much contemporary human rights theorists continue to understand human rights in terms of individual agency and self-determination. To grasp the immensely original contribution Arendt makes to the human rights discussion, it is necessary to grasp the degree to which contemporary human rights theorists continue to think within the framework of Hobbes and Rousseau. Here I turn again to Michael Ignatieff's *Human Rights as Politics and Idolatry* as well as to the recent work of Jürgen Habermas on human rights in *Between Facts and Norms.*

In *Human Rights as Politics and Idolatry,* Ignatieff admiringly points out that the 1948 Universal Declaration of Human Rights "represented a return by the European tradition to its natural law heritage, a return intended to restore *agency,* to give individuals the civic courage to stand up when the state ordered them to do wrong."[25] Ignatieff elaborates on this claim: "Human rights is a language of individual empowerment, and empowerment for individuals is desirable because when individuals have agency, they can protect themselves against injustice. Equally, when individuals have agency, they can define for themselves what they wish to live and die for."[26] This last statement reveals Ignatieff's proximity to the modern tradition in which rights are understood in terms of individual agency and self-determination: as autonomous agents, individuals have the right to determine for themselves how they wish to live their own lives. At the same time, power is understood in terms of individual empowerment: human rights language both empowers (Hobbes) and restrains (Rousseau) individual agency.

Ignatieff points out that in this schema, human freedom or liberty is understood as "negative," referring to Isaiah Berlin's notion of the liberty inherent in human rights: "By agency, I mean more or less what Isaiah Berlin meant by 'negative liberty,' the capacity of each individual to achieve rational intentions without let or hindrance. By rational, I do not necessarily mean sensible or estimable, merely those intentions that do not involve obvious harm to other human beings."[27] Significantly, by linking rational intentions with negative liberty, Ignatieff implicitly continues to embrace the modern theoretical framework of human rights; namely, that framework in which human rights belong to self-interested rational agents for whom freedom is understood in terms of a general though not total lack of hindrance in pursuing self-interested goals. While Ignatieff does not follow Hobbes in reducing the right to power, he continues to embrace the

Hobbesian framework: human rights belong to self-interested rational agents whose liberty is understood negatively.

At the end of his analysis, Ignatieff gives one final reason for the "secular defense of human right"—namely, human suffering and our natural desire to avoid it, which he argues is "a fact about us as a species."[28] This naturalistic desire to avoid pain and suffering coupled with our "limited capacity for empathy" in recognizing the pain of others provides us with a minimal normative basis for empowering individuals with civil and political rights. With this, Ignatieff brings together the fundamental arguments of Hobbes and Rousseau: a natural desire to avoid pain or even death for oneself is conjoined with a natural capacity for pity toward others. Akin to Rousseau, he equates the latter with a free conscience. Oddly, after a long critique of human rights idolatry, Ignatieff ends up justifying human rights on the basis of individual conscience. While Ignatieff is convinced that this negative basis for individual rights has the advantage of not justifying "inhumanity on foundational grounds," he overlooks precisely how limited our natural empathy actually is. Likewise, he overlooks how this limitation itself serves as a basis for all forms of cruelty toward those outside the boundaries of our empathy or pity.

No one has more tellingly reflected upon the limitations of pity than Hannah Arendt. Indeed, Arendt would argue that Ignatieff's location of human rights in individual agency is what leads him ultimately to appeal to something like a limited sense of empathy, which, she argues, is inseparable from the private behavior of individuals guided by customs and mores. Here she is influenced once again by Montesquieu, who argues in *The Spirit of the Laws* that a nation is held together by laws and customs: "Laws govern the actions of the citizen and customs govern the actions of man." Laws, she argues, "establish the realm of public political life and customs establish the realm of society." And she agrees with Montesquieu that nations begin to collapse when law is undermined and the only barriers against political evil are patterns of moral behavior that are no more than customs and habits: "So long as [customs and traditions] are intact men as private individuals continue to behave according to certain patterns of morality. But this morality has lost its foundation. Tradition can be trusted to prevent the worst only for a limited time. Every incident can destroy customs and morality which no longer have their foundation in lawfulness; every contingency must threaten a society which is no longer guaranteed by citizens" (*EU,* 315). Arendt agrees with Montesquieu that there are political dangers when a political body is held together by the "binding force of morality" alone, which in Ignatieff's case is the sense of empathy tied to a free conscience: "We know only too well the alarming speed with which they [habits and customs] are unlearned and forgotten when new circumstances demand a change in manners and patterns of behavior" (*LMT,* 15). Thus, Eichmann's conscience "functioned in the expected way for about four weeks, whereupon it began to function the other way around" (*EJ,* 95). For Arendt, human rights belong to public actors and not private individuals. They therefore cannot have their basis in moral sentiments. That Ignatieff ends up appealing to a moral sentiment as the basis of human rights reveals just how much his theoretical framework is oriented by a notion of the private self-interested rational agent.

Moreover, Arendt would see Ignatieff's appeal to a limited empathy as itself evidence of just how much even our most prominent and thoughtful political theorists are mired in banal truisms and wishful illusions. In her essay "The Eggs Speak Up," she recounts a story told by Ignazio Silone in which a former revolutionary came to see him and in a fit of fervor and intensity said, "One always should act towards others as one wants them to act towards himself."[29] (This is the ultimately the basis upon which Ignatieff stakes his claim concerning empathy.) For Arendt, this story reveals the predicament of those who escaped from "totalitarian hell." They seem to be left with nothing but the "very truisms, moral or otherwise, from which they escaped twenty or thirty years ago—escaped for the very good reason that they had found them no longer sufficient either to explain the world we live in or to offer a guide for action within it" (*EU,* 280). What is frightening for Arendt is that these moral truisms replace the hell of totalitarianism with a kind of "banal philistinism." Again agreeing with Montesquieu, she argues that mores depend on an intact tradition that keeps them alive and that it is impossible "to pretend that [the past] is alive in the sense that it is in our power to return to it, that all we have to do is to listen to the voices of the dead" (*OT,* 282). The loss of such an intact living tradition in the twentieth century transforms morals into impotent truisms, clichés, and banalities that can in no way ward off the dangers of another totalitarian hell and certainly cannot be appealed to as the basis of human rights.

Arendt's analysis of the sentiment of pity in the context of the French Revolution is instructive in understanding why Ignatieff's appeal to the sentiment of empathy with regard to the other's suffering, similar to Rousseau's appeal to pity, not only bases human rights on a moral truism but also can be politically dangerous or may even end in political terror: "Whatever theoretically the explanations and consequences of Rousseau's teachings might be, the point of the matter is that the actual experiences underlying Rousseau's selflessness and Robespierre's 'terror of virtue' cannot be understood without taking into account the critical role compassion has come to play in the minds and hearts of those who prepared and of those who acted in the course of the French Revolution" (*OR,* 79). As we saw above, pity for Rousseau is a "natural virtue" that offers a natural ethical foundations to society, just as Ignatieff's empathy is a "natural fact" of the human species. Arendt argues, *pace* Bernard Flynn, that a society founded on "natural virtue" is terrifying because it reduces the political to the natural. Such a reduction, she argues, is inherently violent. This is particularly true of the sentiment of pity. Pity, as Nietzsche took pains to show, is a sentiment and not a passion. It is reactive rather than active; its passivity makes action impossible. Moreover, like any sentiment, pity is self-involved. Nowhere is this more striking than in Rousseau's *Emile*. Pity is the sentiment that protects the *amour de soi* from becoming corrupted. Through pity, Emile is able to look equitably on society, happy with who he is and relieved not to be one of the prisoners and beggars he sees.

The problem for Arendt is that the distance and self-involvement that characterize the sentiment of pity are such that pity leads to the glorification of its own cause. Bound to no singular, it is without limit: "Since the days of the French Revolution, it has been the boundlessness of their sentiments that made revolutionaries so curiously insensitive to reality in general and to the reality of persons in particular, whom they felt no compunctions in sacrificing to their 'principles' or to the course of history, or to the cause of revolution as such" (90). This boundlessness of ethical sentiment is such that "all must be permitted" to those who act in the revolutionary direction: "The lawlessness of the 'all is permitted' sprang here still from the sentiments of the heart whose very boundlessness helped in the unleashing of a stream of boundless violence" (92).

The violence of natural virtue is accomplished by the reduction of the political to the natural, opening the door to the domination of the social. In other words, the sentiment of pity identifies with the unfortunate, *les misérables*. Arendt argues that because *les misérables* were under the sway of necessity, necessity entered into the realm of the political. The revolution became interested in liberation from misfortune and not the liberation from tyranny through the establishment of justice. From this followed the terror and violence of Robespierre: "It was determined by the exigencies of liberation not from tyranny but from necessity, and it was actuated by the limitless immensity of both the people's misery and the pity this misery inspired." The terror and violence were produced by the reduction to the realm of the natural, which carries the reduction to the singularity of voice. When we are hungry we cry out in one voice; we cry for bread (94). Politically, this spells trouble. The word "people" loses the meaning of plurality. There is, instead, the demand that the social body be one, that it speak with one miserable voice. And it is this demand for unanimity that characterizes the terror of sentimentality.

Thus, for Arendt, the sentiment of pity or empathy has an inherent cruelty. Because it can be enjoyed for its own sake, it is a sentiment, not a passion. It is boundless, bound to no singular; it is endless in its reach. It is without limit and therefore without law. The sentiment of pity is ultimately formless; the nation becomes an ocean with many moods and swells that render impossible the constitution of a political space: "It was indeed the ocean of misery and the ocean-like sentiments it aroused that combined to drown the foundations of freedom" (94). Pity or empathy as a political principle is not impartial, which is why Ignatieff correctly refers to it as a limited sense; it identifies itself only with misfortune and with those who are suffering: "Therefore it has just as much vested interest in the existence of the unhappy as thirst for power has a vested interest in the existence of the weak" (89). Finally, it reduces the political to the natural—pity or empathy, as both Rousseau and Ignatieff argue, is an instinctual reaction to the suffering of others. Nowhere is this more clearly seen than in the trial of Eichmann, who testified of his distress at the suffering of those he saw in the camps and more than once spoke of how he sought to alleviate this suffering. Arendt points out that his distress at the suffering of others reveals that pity and distress are instinctive natural reactions that affect human beings when they are in the presence of suffering. It was easy enough for Eichmann to turn these instincts around and direct them toward himself: "So that instead of saying: what horrible things I did to people, the murderers would be able to say, 'What horrible things I had to watch in the pursuance of my duties, how heavily the task weighed upon my shoulders'" (*EJ*, 106).

While Arendt rejects natural sentiment as a basis for human rights, nevertheless she argues for an affective dimension to the principle of *initium*. Here she is indebted to Montesquieu. She insists that the principle of *initium* is accompanied by the pleasure of appearance. For Arendt, this pleasure is a properly political affection because it animates our initial appearance in a plurality with others.

While Habermas does not posit an affective basis for human rights, his views are important to our discussion. He, like Ignatieff, continues the modern formulation of human rights by positing at the center of human rights the work of negative freedom. Negative freedom is rooted first in individual agency and then generalized as the sovereign will-formation of a people. In his analysis of the negative freedom at work in the Declaration of the Rights of Man and the Citizen, Habermas approvingly refers to Article 4: "Political liberty consists in the power of doing whatever does not injure another. The exercise of the natural rights of every man has no limits other than those which are necessary to secure to every other man the free exercise of the same rights; and these limits are determinable only by the law."[30] Article 4 explicitly delineates natural rights as rights that protect the individual from unpermitted intrusions on his or her freedom, life, and property. Because they are inalienable, there are no limits to these natural rights other than the security of every other individual's freedom, and these limits are established by law. Habermas says, "The concept of individual rights plays a central role in the modern understanding of law. It corresponds to the concept of liberty or individual freedom of action: rights ('subjective rights' in German) fix the limits within which a subject is entitled to freely exercise her will. More specifically they define the same liberties for all individuals or legal persons understood as bearers of rights."[31] Habermas understands rights as individual liberties that need to be supplemented by rights "of a *different* kind, rights of citizenship that are geared no longer to rational choice but to autonomy in the Kantian sense."[32] Private subjects with private interests must "drop the role of the private subject and assume along with their role of citizen, the perspective of members of a freely associated legal community, in which an agreement on the normative principles for regulating social life either has already been secured through tradition or can be brought about deliberatively in accordance with normatively recognized procedures."[33]

Habermas finds himself in a theoretical muddle. On the one hand, he embraces the modern understanding of human rights located in the wills of rights-bearing individuals, while on the other hand he argues that the individual will cannot itself be grasped theoretically in terms of a moral subject: "The traces of modern natural-law normativism thus get lost in a trilemma: neither in the teleology of history nor in the constitution of the human species can we find the content of practical reason, once its philosophical foundation in the knowing subject has been shattered."[34] Not surprisingly, Habermas moves from the modern subject to rational communication as the site that legitimizes human rights. Yet he is not able to relinquish a notion of rights based on the autonomous will-formation of a people: "The co-originality of private and public autonomy first reveals itself when we decipher, in discourse-theoretic terms, the motif of self-legislation according to which the addressees of law are simultaneously the author of their rights. The substance of human rights then resides in the formal conditions for the legal institutionalization of those discursive processes of opinion- and-will-formation in which the sovereignty of the people assumes a binding character."[35] He asks, "What grounds the legitimacy of rules that can be changed at any time by the political lawgiver?" He replies, "This question becomes especially acute in pluralistic societies in which comprehensive worldviews and collectively binding ethics have disintegrated, societies in which the surviving post-traditional morality of conscience no longer supplies a substitute for the natural law that was once grounded in religion and metaphysics."[36] Habermas's answer is striking in its simplicity: "Discourse theory answers this question with a simple, and at first glance unlikely, answer: democratic procedure makes it possible for issues and contributions, information and reason to float freely; it secures a discursive character for political will-formation."[37] For Habermas, the modern legal order draws its legitimacy from self-determination, which is constituted not "by way of a social contract but on the basis of a discursively achieved agreement."[38]

Habermas wants to find a basis for legal normativity and human rights, and he finds it in the will-formation of a sovereign people. The fact that this will-formation is constituted on the model of communicative discourse and not generalized reason does not make any difference: the people's will becomes the basis of legitimate and illegitimate law. Seen from an Arendtian point of view, Habermas has not moved very far from Rousseau: the people's will is still the unanimous will of a sovereign nation. While it is true that Habermas distinguishes popular sovereignty, which integrates the nation on the basis of descent, shared tradition, and common language, from state sovereignty, whereby the unity of the nation is achieved through the political identity of the citizen, nevertheless he collapses the distinction in the same way Rousseau does. (It is telling that he refers to the French Revolution when invoking state sovereignty.) Discursive agreement continues to be thought of as the sovereignty of a people who remain homogeneous in their discursive agreement. The rationality at work in Habermas's thought continues to be the rationality of a willful "people." The nightmare of a *Volk* is not banished.

Nowhere is this more apparent than in his postscript to *Between Facts and Norms*. Here he argues that the "solidarity" characteristic of discursive agreement is achieved by "stabilizing behavioral expectation." While this could be read as a regularized procedure toward the creation of law, Habermas's discussion of the foreigner and the alien reveals that for him the "behavioral expectation" extends beyond procedural legalism, reaching into the very sense of something like the behavior of a "people." This is clear when he argues that "contrary to what the model of a legally regulated moral division of labor suggests, the social boundaries of a political community do not have just a functional meaning. Rather, they regulate one's belonging to a historical community of shared destiny and a political form of life that is constitutive for the citizen's very identity."[39] While it might seem again that Habermas is pointing to a notion of state sovereignty in which the identity of the citizen is strictly *political*—that is, *legal*—this is not the case. Habermas argues that aliens, displaced foreigners, and stateless persons "have at least approached the status of citizens because of the human rights meaning of these basic rights; these groups enjoy the same legal protection and have, according to the letter of the law, similar duties and entitlements."[40] Political will-formation continues to be thought of in terms of homogeneity of a people. "Citizenship is an answer to the question 'Who am I' and 'What should I do' when posed in the public sphere. Membership in a political community grounds special duties behind which stands a patriotic identification. This kind of loyalty reaches beyond the validity of institutionally prescribed legal duties."[41] Like Rousseau, Habermas ends up collapsing state sovereignty into popular sovereignty—that is, the sovereignty of a people who have special obligations to the foreigner and the immigrant but whose obligations do not extend to the granting of rights. Rights, for Habermas, remain the rights of citizens, and the citizens are again nationals whose behavioral expectations, including patriotic loyalty and identification, constitute the basis of rightful membership. Although Habermas wishes to avoid this collapse, in Arendt's view, it was inevitable. Basing the legitimacy of right in political will-formation, itself based on the model of the autonomous rational will of an individual, continues the modern identification of human rights with the homogeneity of a sovereign people. Nothing changes when the homogeneity is obtained through rational discursive agreement: the alien, the foreigner, and the immigrant are still not rightful members of the discourse.

THE PRINCIPLE OF *INITIUM:* RETHINKING FREEDOM, POWER, AND ACTION

Arendt's explicit formulation of the right to have rights is in large part carried out not through a critique of the "fiction of human nature" but through a critique of the notion of sovereignty. She claims that a notion of sovereignty as much as a certain conception of the subject underpins the modern formulation of human rights. Arendt's analysis suggests that it might be possible to reformulate the notion of human rights by thinking a notion of power and right that is not tied to the notions of sovereignty and national citizenship. Her rejection of sovereign power has its basis in her rejection of the dominant philosophical understanding of freedom, which from Augustine onward has largely understood freedom as located in a subjective will. Her understanding of freedom, in contrast, is political, located not in the "I will" but in the "I am able." Indebted to Aristotle, she argues that freedom is always the freedom to move and is by definition worldly. She argues that the experience of worldly freedom is the condition for a notion of subjective freedom (a notion that Arendt herself rejects): "Hence, in spite of the great influence the concept of an inner, nonpolitical freedom has exerted upon the tradition of thought, it seems safe to say that man would know nothing of inner freedom if he had not first experienced a condition of being free as a worldly tangible reality" (*BPF,* 148). Freedom, she argues, is "experienced in the process of acting and nothing else." This capacity to act and move must be understood as the capacity to begin: "The Greek word *archein* which covers beginning, leading, ruling, that is, the outstanding qualities of the free man, bears witness to an experience in which being free and the capacity to begin something new coincided" (166). From the outset, Arendt's understanding of freedom is inseparable from power, the ability to begin. The "I am able" must be understood as the ability to act in a public space, to move in a space of freedom with others.

According to Arendt, power must always be said in the plural. For power to exist there must be other centers of power: "Power comes into being only if and when men join themselves together for the purpose of action, and it will disappear when, for whatever reason, they disperse and desert one another" (*OR,* 175). Domination, in contrast, is the loss of power that occurs only where there is a central ruling power. The notion of sovereignty can denote strength, but it can never denote power. The principle of federalism, she argues, illuminates the point. The establishment of the Union, as Madison and Jefferson understood, did not take away from the power of the states but instead provided a new source of power. She argues that if the individual states had not existed, the Union would have had to erect them in order to have the power it did. This again suggests that power is generated by power and that to be powerful one must be in relation to other powers. Here we see Arendt's rejection of Rousseau's identification of sovereignty with power. Arendt argues instead that action demands a plurality of actors. Further, power must be understand as the only "human attribute which applies solely to the worldly in-between space by which actors are mutually related" (175).

Power therefore denotes not only the ability to act but action in concert with others.[42] Arendt insists that "the power structure itself precedes and outlast all aims, so that power, far from being the means to an end, is actually the very condition enabling a group of people to think and act in the means-ends category" (151). This is why Arendt argues that full-blown terror, resulting in the complete atomization of the political, is caused by the presence of an absolute violence without the presence of power. Power, which is present only when people act in concert, has completely disappeared. Montesquieu already saw this problem: terror is ultimately potent and self-destructive because it fears any and all organization, even turning against those in its own forces who might organize. Arendt argues that the difference between totalitarianism and tyranny is that the former turns even against the power of its friends (*OV*, 55).

How then can we think legitimate power? Arendt's understanding of a noncentralized nonsovereign power that is synonymous with public freedom and action allows us to rethink a legitimate principle of power. Here we must understand that because the modern understanding of law is rooted in Hobbes, it is contractual. Furthermore, the contractual understanding of law is inseparable from an understanding of power as sovereign. Arendt argues, however, that the contract is not the exclusive foundation for the law. In contradistinction to the contract, the law can be founded on the mutual compact. She points to compacts, such as the Mayflower Compact, that were made prior to the Revolution and that were made with no reference to prince or king. The principle of the compact (or covenant) is the claim to power without the further claim to sovereignty. The principle was "neither expansion nor conquest but the further combination of powers" (*OR*, 168). The compact, Arendt argues, understands the political bond in the old Roman sense of alliance: "Such an alliance gathers together the isolated strength of the allied partners and binds them into a new power structure by virtue of 'free and sincere promises'" (170). Again, she argues, this is very different from the contract wherein an individual person resigns his "power to some higher authority and consents to be ruled in exchange for a reasonable protection of his life and property" (169). The difference between an act of covenant and an act of consent is that the first is based on an increase of power through the recognition of others inspired by the principle of plurality, while the second is based on the surrender of power in the recognition of sovereignty that is inspired by the principle of unanimity.

The notion of law that emerges out of the mutual covenant is one that understands the law as neither sovereign nor dominating, neither commandment nor imposed standard. Rather, following Montesquieu's insight, Arendt suggests that the law must be understood as regulator of different domains of power. Here there is a way to think multiplicity of power with rule. Arendt again looks to Montesquieu, for whom the law "never lost is original 'spatial significance' altogether, namely, 'the notion of a range or province within which defined power may be legitimately exercised'" (186–187). And since the laws are no more than the relations that exist and preserve different realms of power and are therefore relative by definition, Arendt argues that Montesquieu "needed no absolute source of authority and could describe the 'spirit of the laws' without ever posing the troublesome question of their absolute validity" (189).

Yet Arendt does not relinquish the problem of legitimate principles of power, and it is this that moves her from an analysis of power to the formulation of a principle of humanity that is the basis for distinguishing legitimate and illegitimate forms of power. She accomplishes this through an analysis of the event of natality, drawing on the triple meaning of the Greek word *arche,* which, she argues, conveys the sense of principle, beginning, and common ground. The principle of action, she argues, lies in its beginning, and it is this principle of *initium* that provides the inspiration for power and action. If it is the case that power is synonymous with action and freedom, and further, if all three terms denote the appearance of an actor among a plurality of actors in a space of freedom, then the principle that inspires power and political action and that lends action its legitimacy is the principle of publicness: "Because of its inherent tendency to disclose the agent together with the act, action needs for its full appearance the shining brightness we once called glory, and which is possible only in the public realm" (180). Insofar as action and power are synonymous terms for Arendt, we can substitute power for action in the above passage. Her analysis suggests that legitimate power is precisely that power that allows the actor to appear in a public space with others. This is the principle that ought to inspire the constitution of the political space and all activities carried out therein: it demands that the divisions of power be such that all actors are able to appear and act. The principle of publicness demands that all positive civil laws constitute and regulate the divisions of power in such a way that all the actors are empowered.

This principle also permits the establishment of a notion of rights that has another basis than an understanding of human beings as sovereign subjects endowed with inalienable rights. Arendt is able to reject the fiction of human nature and still think the inalienable right of the actor who in order to act must be able to appear in a public space of freedom. Though stripped of state and home, the actor must not be stripped of this fundamental right to be able to appear, because the first act, the act of beginning itself—the event of natality—contains both the beginning and its principle within itself.

The event of natality that carries within it the principle of publicness, when restated as the law of humanity (understood as the appearance of the actor among a plurality of actors in a public space of freedom), demands that the actor have the right to appear, or, as Arendt so succinctly puts it, the right to have rights. This right is not predicated on a metaphysical understanding of the human being as having a nature; instead, it is predicated on the fundamental event of human existence—natality. To be born is to appear on the globe. This is the law of humanity that Kant articulated in his essay *Perpetual Peace* and that Arendt quoted so approvingly in her Kant lectures: "Humans have [the right of appearance] by virtue of their common possession of the earth, whereas on a globe, they cannot infinitely disperse and hence must finally tolerate the presence of each other. The common right to the face of the earth . . . belongs to human beings generally" (*KPP,* 75). The right to have rights is inspired by a new principle of humanity, the principle of publicness that demands that each actor by virtue of the event of natality itself has the right to temporary sojourn on the face of the earth.

The principle of plurality is inseparable from the principle of publicness. Perhaps it is even tautological to posit the second. However, it clarifies Arendt's understanding of "publicness," answering Castoriadis's criticism of Arendt that "Stalin appeared." From an Arendtian perspective, Stalin did not appear. For Arendt, "appearance" inherently carries the condition of plurality. As she argues in *The Human Condition,* appearance means to be seen and heard by others (*HC,* 50). The fundamental condition of being human, she argues, "is that men and not man inhabit the world" (7). In other words, "plurality is the law of the earth." Being and appearing are coincident, and appearance requires a spectator: "Nothing and nobody exists in this world whose very being does not presuppose a *spectator.* In other words, nothing that is, insofar as it appears, exists in the singular; everything that is is meant to be perceived by somebody" (*LMT,* 19, emphasis in the original text). For Arendt, "world" is constituted only through a plurality of perspectives. Since the fundamental event, the beginning, carries its principle within it, the event of natality, the event of appearance, carries the principle of plurality—with its inherent publicness—within it. To violate this principle is a "crime against humanity." This is why Arendt argues at the end of her book on Eichmann that genocide is a crime against humanity.

> It was when the Nazi regime declared that the German people not only were unwilling to have any Jews in Germany but wished to make the entire Jewish people disappear from the face of the earth that the new crime, the crime against humanity—in the sense of a crime "against the human status" or against the very nature of mankind—appeared. Expulsion and genocide, though both international offenses, must remain distinct; the former is an offense against fellow-nations, whereas the latter is an attack upon human diversity as such, that is, upon a characteristic of the human status without which the very words "mankind" or "humanity" would be devoid of meaning. (*EJ,* 268–269)

This passage makes clear that the new principle of humanity that Arendt called for as early as the summer of 1951 includes the principle of plurality and its publicness. At this point, we see that Arendt has answered Benhabib's criticism that she offers no philosophical justification for the category of crimes against humanity. Crimes against humanity are crimes that attempt to eradicate plurality from the face of the earth. The attempt to physically exterminate the Jewish people from the face of the earth was an attempt to eradicate the plurality inherent in the principle of humanity. Genocide, the systematic attempt to eradicate plurality, is a crime against humanity. This is why Arendt argues that Eichmann, having been found guilty of this crime against humanity, was no longer fit to live among human beings.

Moreover, for Arendt, "appearance" carries the notion of significant speech and action. It is not enough to give someone a place of refuge; instead, the principle of plurality and its inherent publicity demands that there be a public place where one is truly seen and heard. Otherwise, she argues, one is simply the fool or idiot, one who speaks or acts without significance, which is just another kind of invisibility. Moreover, one must be able to initiate human action in concert with others. It is our "freedom to begin something new and unexpected that was not there before" (*BPF,* 151). For Arendt, significant speech and action, as well as the capacity to begin something new, can occur only in a political space. Thus, the right to have rights, which is established through the principle of *initium* restated as the principles of publicness and plurality, is the right to belong to a political space. These principles carry the rights of freedom of expression and association, I would argue. In this context, Arendt's approving reference to Pericles' funeral oration receives added import: "Wherever you go, you will be a *polis*" (*HC,* 198). Understood on the basis of the right to have rights, this means that wherever one goes, one has the right to belong to a political space where significant speech and action and the capacity to initiate are possible. This is because each human being is an appearance who requires a public space in order to truly appear: "To be deprived of it means to be deprived of reality, which, humanly and politically speaking, is the same as appearance" (199).

Significantly, for Arendt, the right to have rights is more than merely a juridical right. For her, the right to have rights is a fundamental political right; it is the right to belong significantly to a political space. Arendt's distinction in *The Human Condition* between the law as the wall around the political space and the political space itself is helpful here. The public space is a space of significant speech and action, while the law is that which regulates such action. The right to have rights has as much to do with political representation and the possibility of political action as it does with formal equality under the law. The mistake the Jewish people made in the eighteenth and nineteenth centuries, she argues, was to put too much faith in juridical rights—namely, equality before the law—while being completely unconcerned with rights that would have gained them entry into the public space through political representation. Such lack of political concern, itself a consequence of lacking political rights, was disastrous.

By providing a new principle of humanity, Arendt is able to distinguish between legitimate and illegitimate shapes of power and political action without having to invoke the ethical. Power, which is synonymous with acting politically with others, must be inspired by the categorical imperative of the political: the principle of plurality provides us with a new law of humanity, demanding that each actor, by virtue of the event of natality itself, has the right to appear with others, the right to act and speak within the political space.

Arendt's understanding of the originary right to belong to a public space is close to Amartya Sen's articulation of rights as capabilities that are features of persons rather than characteristics of goods. Sen argues that human rights articulate positive freedoms; namely, what a person can and should be able to do. The freedom of a person must be understood in terms of the functioning of the person: "capability to function reflects what a person *can* do."[43] Thus, he argues, "Concern with positive freedoms leads directly to valuing people's capabilities and instrumentally to valuing things that enhance these capabilities. The notion of capabilities relates closely to the functioning of a person. This has to be contrasted with the ownership of goods, the characteristics of goods owned, and the utilities generated."[44]

Arendt's notion of freedom is very similar. She looks to Montesquieu's understanding of the constitution of political freedom, arguing that "for him the word 'constitution' in this context has lost all connotations of being a negative, a limitation and negation of power; on the contrary, the word means that the 'grand temple of federal liberty' must be based on the foundation and correct distribution of power" (*OR*, 150). She further argues that Montesquieu "had maintained that power and freedom belonged together, that conceptually speaking, political freedom did not reside in the I-will but in the I-can, and that therefore the political realm must be construed and constituted in a way in which power and freedom would be combined" (150). Here Arendt rethinks the Hobbesian relation between power and right. Rather than seeing right and power as oppositional terms in which the first is most often reduced to the second, Arendt is arguing for a notion of right that is inseparable from political power, the latter understood in terms of belonging to a public space that guarantees not only juridical rights but also political representation such that significant speech and action is possible.[45] Here Arendt reveals her distance from both Ignatieff's and Habermas's understandings of freedom as negative. For Arendt, freedom as the "I am able" is positive and requires a public space of action. Making a distinction between liberty and freedom, Arendt argues that "all these liberties, to which we might add our own claims to be free from want and fear, are of course essentially negative; they are the results of liberation, but they are by no means the actual content of freedom, which, as we shall see later, is participation in public affairs, or admission to the public realm" (32). Liberation is the condition of freedom, but some addition, namely the public space, is necessary for the exercise of freedom. Speaking of the eighteenth-century revolutionaries, Arendt argues that "the acts and deeds which liberation demanded from them threw them into the public business, where, intentionally or more often unexpectedly, they began to constitute that space of appearances where freedom can unfold its charms and become a visible, tangible reality" (33). The pleasures and charms of freedom require a worldly space of appearance.

Arendt's notion of the right to have rights as an originary positive freedom is inseparable from the obligation of common responsibility to enact this right. To use Sen's example, we are obligated to say "yes" not only to the person who has been robbed (whose negative freedom has been violated); we must also say "yes" to those who have suffered loss from flood, drought, or political displacement (even though none of their negative freedoms have been violated).[46] Responsibility for others is at the very heart of freedom.

There is, however, an important difference between Arendt and Sen on the issue of these positive freedoms. For Sen, positive freedoms are always embodied. While he agrees with Arendt that freedom must be understood as the "I am able," he insists on the consequence of this understanding. Freedom is always embodied: "Depending on our body size, metabolism, temperament, social conditions, etc., the translation of resources into the ability to do things does vary substantially from person to person and from community to community, and to ignore that is to miss out on an important general dimension of moral concern."[47]

Henry Shue's notion of "basic rights" also adds to Arendt's account of positive freedom. Shue uses the example of the "right to free association," arguing that it is not enough to have the right; one must be able to enjoy it. The last requires the accompanying right to the fulfillment of vital needs such as food, shelter, health care, and education. Only with these basic rights fulfilled are we able to enjoy civil and political liberties. Basic rights, having to do with our vital needs, "specify the line beneath which no one is to be allowed to sink." They "are a shield for the defenseless against at least some of the more devastating and more common of life's threats, which include . . . loss of security and loss of subsistence."[48] It is striking that

Arendt ignores this dimension of freedom. While she writes eloquently and at length on freedom as the freedom to move, she seems to forget entirely that this movement is always an embodied movement. Such forgetfulness allows her to make too strict a distinction between the political and the social.

HUMAN RIGHTS AND THE PUBLIC USE OF REASON

In her essay "Truth and Politics," Arendt states:

> Political thought is representative. I form an opinion by considering a given issue from different viewpoints, by making present to my mind the standpoints of those who are absent; that is, I present them. This process of representation does not blindly adopt the actual views of those who stand somewhere else and hence look upon the world from a different perspective; this is a question neither of empathy . . . nor of counting noses and joining a majority, but of being and thinking in my own identity where actually I am not. (*BPF*, 241)

Representation is the activity of making present absent standpoints; it is the work of the political imagination, requiring the enlarged mentality of Kant's *sensus communis*. The question I want to explore in the final section of this chapter is how the temporality of the deflected present that characterizes the temporality of natality is at work in the enlarged mentality and how this enables us to grasp Arendt's understanding of the representation of the nation or, more precisely, the representation of the "we." If we think the activity of making present absent standpoints through the temporality of natality, it is impossible to read Arendt as offering a notion of the public space that is consensual and collusive based on either the shared standpoint of the community and its traditions or anything like the generalized interest of reason, the latter characteristic of the Habermasian public space. The activity of making present absent standpoints disallows anything like a sovereign or national "we" as constitutive of the public space. Arendt insists that the public space must be understood as "sameness in utter diversity" and that only in this way "can worldly reality truly and reliably appear" (*HC*, 57).

The enlarged mentality for Arendt is not empathic, nor does it involve the actual standpoints of others. It does not indulge the fiction that one can assimilate the other's standpoint as if it were possible to make oneself at home elsewhere. At the same time, emphasizing the possible standpoints of others, Arendt propounds a notion of the public space that is always "potentially public, open to all sides" (*KPP,* 43). She thereby avoids positing anything like a communitarian model, in which action is reducible to a unanimous general will rooted in a shared history, beliefs, and identity. The enlarged mentality adopts the "general standpoint," which, Arendt argues, "is not the generality of the concept—it is, on the contrary, closely connected with the particular conditions of the standpoints one has to go through in order to arrive at one's own 'general standpoint'" (43–44). Arendt is clear that the general standpoint is not the common or general view that is held by all in the *sensus communis,* as though it were akin to Rousseau's general will.

Indeed, Arendt's distance from Rousseau on this point could not be greater. For Rousseau, the general will is divested of any and all particular interests; it is one, indivisible, and unified. For Arendt, the general standpoint is made up of a plurality of standpoints which together constitute a common world. The more standpoints, the richer and more nuanced the world:

> The more peoples there are in the world who stand in some particular relationship with one another, the more world there is to form between them, and the larger and richer that world will be. The more standpoints there are within any given nation from which to view the same world that shelters and presents itself equally to all, the more significant and open to the world that nation will be. (*PP,* 176)

For her, Rousseau's unanimous general will constitutes the end of the world: "If matters in a nation were to come to a point where everyone saw and understood everything from the same perspective, living in total unanimity with one another, the world would have come to an end in a historical-political sense" (ibid.). If Rousseau is the exemplary thinker of the modern nation-state, then the seeds of its decline are found in the notion of the general will. Arendt's distance from Rousseau's general will is all the more significant when we recall that for her, totalitarianism is characterized by a thinking that "everything is possible," the latter coming about only when utilitarian motive and self-interest (characterized by "everything is permitted") no longer hold sway in the political realm. Far from denigrating self-interest and utilitarian motive in the constitution of a common political world, Arendt's reflections on totalitarianism suggest that political terror begins when such interest is abandoned. This is why, she points out approvingly, German Marxists who identified with class interests were among the strongest resistors to Nazi ideology.

Here we must recall that Arendt begins to question the modern formulation of human rights on the basis of the anomalous and discriminatory legal status assigned to migrant and refugee populations, who find that they are outside the borders of the nation and thus on the other side of the law. Arendt's critique is that the modern notion of rights cannot accommodate or address the minorities and refugees that are expulsed from the nation's boundaries. With this in mind, we can understand the significance of her insistence that the right to have rights includes the three fundamental rights found in Kant's *Perpetual Peace:* the right to visit foreign lands, the right to hospitality, and the right to temporary sojourn. These fundamental rights, all of which emphasize the rightful status of the refugee and the migrant, are thought on the basis of a political imagination that does not require that the individual belong to a national will and its claims to the continuity of an authentic past.

In Arendt's reading of Kafka, Benjamin, and Heidegger, the force of the past is an anteriority that continually introduces the foreign into the present. For Arendt, thinking political representation and the enlarged mentality through the temporality of natality creates a temporal break in the process of representation that provokes a crisis in the activity of signifying the public space. We then have a contested public space where the "nation's people" must be thought in terms of what Homi Bhaba calls a "double temporality": on the one hand, through the *inter-esse* constituted in and through the web of sedimented histories and enacted stories that are based on traditions (which Bhaba calls "pedagogical temporality"), and on the other hand, through a plurality of actors engaged in the activity of resignification, challenging any original presence of a nation's people (which Bhaba calls "performative temporality").[49] Bhaba fails to understand how close his thinking is to Arendt's notion of the public space as a space of double temporality; instead, he reads Arendt's notion of mimesis as the mere repetition of tradition. Referring explicitly to the Arendtian public space, Bhaba asks "What is temporal in the mode of existence of the political? Here Arendt resorts to a form of repetition to resolve the ambivalence of her argument. The 'reification' of the agent can only occur, she writes, through 'a kind of repetition', the imitation of mimesis, which according to Aristotle prevails in all arts but is actually appropriate in drama."[50] Bhaba argues that Arendt not only understands mimesis as mere repetition but bases this understanding on a notion of the public space as a community constituted through consensual togetherness. I do not agree. Arendt understands mimesis from her understanding of the temporality of natality, which means that for her mimesis is never simple repetition. It allows for the inauguration of the new. Moreover, Arendt explicitly rejects the notion of the public space as constituted by a consensual togetherness. For her, the togetherness characteristic of the public space is constituted through the sameness of utter diversity: dissent and disagreement are integral to it. When Arendt argues that the public space is constituted by a to-getherness "where people are with others and neither for nor against them," she is explicitly referring to the violence of war in which the other is one's enemy. This kind of violence destroys the public space. Moreover, in this passage Arendt also rejects a public space in which we are "for one an-other," which suggests that she is not embracing anything like consensual agreement. Arendt therefore is much closer to Bhaba's position than he is willing to admit. Like Bhaba, her entire thought is motivated by a concern for minorities, refugees, and the politically marginalized. The Arendtian political imagination challenges the imagination of national cultures and their claims to the continuity of an authentic past. The temporality of na-tality renders the public space fundamentally open "on all sides" to the fu-ture.

Moreover, and in contrast to Habermas, who is faced with the prob-lem of how to move from the private reason of rights-bearing individuals to the public reason of lawful citizens, Arendt's refusal to embrace a no-tion of private rights is at the same time a refusal to embrace a notion of private reason. From the start, reason for her is public. Here I want to turn to two questions that Arendt raises in the first volume of *The Life of the Mind*: "What makes us think?" and "Where are we when we think?" Her answer is given in several different places in her writings. Nowhere is it more clearly given than in the essay "Concern with Politics in Recent Eu-ropean Philosophical Thought."[51] Arendt is sympathetic with the Greek assumption that it is wonder that stands at the origin of thinking: it is sheer wonder at what is, as it is, that is the condition for thinking itself. She argues that the nature of the wonder has changed dramatically in the twentieth century. Rather than the wonder at the beauty and order of the world that characterized the ancient Greeks, today the condition for sheer wonder at what is, as it is, is the sheer horror of contemporary political events.

The differences between these two experiences of wonder have signifi-cant consequences for understanding the status of thinking today. The Greek *thaumazein* was wonder at the *kalon*, the beauty of appearances. This wonder extended into the realm of the public. Indeed, the Greek con-cern with virtue, the *kalon k'agathon*, was the concern with the public ap-pearance of the actor: "how he appeared while he was doing" (*LMT*, 131). The concern with greatness or beauty of appearance was at the basis of the desire for immortality, "the precious reward for great deeds and great words" (131). The wonder at the *kalon*, at what is purely what it is, carried the temporality of endurance, of immortality. Thus, prior to the rise of philosophy, the Greek experience of wonder led to the position of the bard "who helped men to immortality," for "the story of things done outlives the act" and "a thinking said walks in immortality if it has been said well" (132). The actors in the public realm needed the poet "lest their beauty go unrecognized." And it is precisely because the light of the public illuminated these deeds and words that the poet was able to recognize their beauty and confer upon them their immortality.

Wonder at the sheer horror of contemporary political events is wonder in the face of darkness. Quoting Heidegger and agreeing entirely with him, Arendt claims that today "the light of the public obscures everything."[52] If wonder is the condition of thinking and is a wonder in the face of obscurity, then the condition of thinking today is a fundamental darkness; it is a thinking that has as its condition the obscurity of the public realm. For Arendt, what makes us think today is the loss of the illuminating light of the public space. It would be a mistake to understand Arendt as arguing for a recovery of the Greek notion of immortality, for a political space that would render possible great words and great deeds. This greatness, the *kalon k'agathon,* has been rendered impossible, given the loss of public illumination. Consequently, the temporality of the political space is no longer that of immortality; immortality is possible only within an abiding tradition. The speechless wonder at sheer horror carries with it the loss of this tradition: the past is a veritable rubble heap around us. What makes us think, therefore, is not only sheer wonder at horror but also the loss of continuity, of an abiding tradition.

This leads Arendt to answer her second question, "Where are we when we think?" Her answer draws inspiration from Husserl and Heidegger, and, more immediately, from Benjamin and Kafka: "What you are then left with is still the past, but a *fragmented* past, which has lost its certainty of evaluation. . . . It is with such fragments from the past, with their sea-change, that I have dealt" (212). Arendt's reading of Kafka's parable "He" elaborates *where* the thinker's place is in relation to this fragmented past. The thinker, "He," stands in the fragment, the gap, between the past and the future. Arendt thinks this gap as the "untimely"; it is the time of the crisis. For Arendt, the notion of the "untimely" calls into question the form of history that reintroduces and always assumes a suprahistorical perspective, a history whose function is to compare the finally reduced diversity of time into a totality fully closed upon itself, a completed development. To think the untimely is to think *krinein:* critical history as Nietzsche understood it in the Second *Untimely Meditation.* Following Heidegger, and anticipating so much of deconstruction and postmodern thought in general, Arendt understands *krinein* as the perspective that distinguishes, separates, and disperses. Thus "He" who moves in the crisis, in the gap between the past and the future, is engaged in the activity of *krinein,* judgment. Here too we begin to understand what Arendt means when she argues that

"thinking is always out of order." The "out of order" is not such that thinking posits a transcendent object. It is always out of order when compared
with the ordinary course of events, but it is never unworldly. Thought is
inherently bound to the world of appearances. Here we can begin to see
why for Arendt the banality of evil is linked to thoughtlessness. If thinking
has to do with a crisis of judgment, a fundamental "out of order" with the
ordinary course of events, then it is the case that thoughtlessness is precisely lack of judgment, the going along with the course of events in which
one is immersed. Certainly Arendt saw that this was the case with Eichmann.

Thus, in her reading of Kant's *Critique of Judgment,* Arendt agrees with
his insistence on a way of thinking "for which it would not be enough to be
in agreement with one's own self, but which consisted of being able to
think in the place of everybody else," a thinking Kant called an "enlarged
mentality" (*eine erweiterte Denkungsart*) (*BPF,* 230). She approvingly points
out that Kant's entire theoretical philosophy depends on the interplay between understanding and sensibility. The thinker cannot leave the cave but
must live among fellow human beings. While the thinker is one who clarifies experience, he or she does not leave experience altogether. Ordinary
human beings can clarify and evaluate their experience, and this sets up
conditions of equality. For Kant, reason is a "general human need," and
there is no distinction between the few and the many.

In Arendt's reading of Kant, thinking is intrinsically anti-authoritarian;
it stands the test of open and free communication, which means that the
more people who participate in it, the better. This is what Kant means by
the "public use of one's reason": thinking requires the company of others.
Arendt points out that this is also what Kant means by political freedom:
"The external power that deprives man of the freedom to communicate his
thought publicly, deprives him at the same time of his freedom to think"
(234). This does not mean that one does not think when one is alone but
only that one's thoughts must be capable of being communicated either
orally or in written form. And, paraphrasing the heroine of Beckett's *Company,* "What an addition to company thinking would make!" Thinking can
be conducted only if accompanied by the freedom to communicate and
exchange thoughts in public, which enables one to enlarge one's mind by
incorporating the insights of others. Arendt argues that thinking not only
depends on the public use of one's reason but also at the same time "feeds
back into public life in its turn by questioning authorities and accepted as-
sumptions" (*KPP,* 38). To think is to judge with a *plurality,* which is the
condition for the public space: "[Thinking] needs the presence of others 'in
whose place' it must think, whose perspectives it must take into consideration, and without whom it never has the opportunity to operate at all"

(*BPP,* 220–221). Here again, Arendt is referring to the general standpoint of Kant's "enlarged mentality." The "place" of these others is precisely the public space.

Discussing Gotthold Lessing's notion of *Selbtsdenken,* Arendt gives a further clue as to how she understands thinking as a worldly activity. *Selbtsdenken* is independent thinking, but it does not mean an isolated individual who looks around the world in order to bring him or herself into harmony with the world by the detour of thought. Thought, she argues, does not arise out of the individual and is not the manifestation of a self. This is still another way to understand the deflective movement of the "He" between past and future: "The individual—whom Lessing would say was created for action, not ratiocination—elects such thought because he discovers in thinking another mode of moving in the world of freedom" (*MDT,* 9). Arendt thinks the proximity of thought and action. Like action, thinking is not engaged in a means-end relation. Thinking has nothing to do with conclusions or results. Like action, it is caught up in beginnings. Here Arendt reveals how much she removes thinking from prescribed categories or standards of judgment; she argues that thinking (judging) is concerned with events and that "what the illuminating event reveals is a beginning in the past which had hitherto been hidden." Thus there is no need to ask how we get from private to public reason, nor is there any need to wring our hands over how we are going to provide moral norms to shore up the laws of our public space. For Arendt, even though the Holocaust exploded our traditional categories of thought and standards of judgment lie around us in a veritable rubble heap, we need not despair:

> Even though we have lost yardsticks by which to measure and rules under which to subsume the particular, a being whose essence is beginning may have enough of origin within himself to understand without preconceived categories and to judge without the set of customary rules which is morality. If the essence of all, and in particular of political, action is to make a new beginning then understanding becomes the other side of action, namely, that form of cognition, distinct from any others, by which acting men . . . eventually can come to terms with what irrevocably happened and be reconciled with what unavoidably exists. (*EU,* 321–322)

Our beginning contains the principle of humanity, which Arendt rethinks as the principle of plurality/publicness, but always in the context of *initium* and natality. This principle of humanity provides us with the necessary "norm" to guide all thinking and action. Thinking is animated by the principle of plurality and publicness, rather than by a private reason that must be generalized (which gives rise to Habermas's perplexity about how we gain rights of citizens from the private rights of individuals). Arendt argues that all along rights have been public. Rights belong to individuals who are *inherently* public and find themselves always in a plurality with others, in a web of relationships from which there is no escape. As we will see in the next chapter, this does not mean that Arendt could be listed among the communitarians, for there is also a principle of uniqueness and singularity at work in the right to have rights.

THREE

The Principle of Givenness:
Appearance, Singularity, and
the Right to Have Rights

> Every man, being created in the singular, is a new beginning
> by virtue of his birth; if Augustine had drawn the consequences
> of these speculations, he would have defined men, not,
> like the Greeks, as mortals, but as "natals."
> Hannah Arendt, *Life of the Mind,*
> vol. 2, *Willing*

> Without a stage-set, man cannot live. The world, society, is
> only too ready to provide another if a person dares to toss
> the natural one, given him at birth, into the lumber room.
> Hannah Arendt, *Rahel Varnhagen*

For Hannah Arendt, our capacity for beginning is the only promise left after the horrifying events of the twentieth century. Augustine's notion of beginning, or natality, is at work throughout Arendt's work after *The The Origins of Totalitarianism,* informing her key concepts of action, freedom, and power. These three notions have as their ontological basis the event of natality: our freedom and power to act are the result of our being born—the *initium.*

Oddly, the political distinctions that Arendt draws from Augustine's understanding of the event of natality seem to eschew all that is usually associated with this event, most notably all that has to do with embodiment. Arendt seems to use Augustine's understanding of the human being as a beginner as the basis for distinguishing the public from the private, freedom from necessity, and action from labor. All these distinctions have as their basis Arendt's foundational distinction between *zoe* (life) and *bios politikos* (political life). Her claim that each human being has an equal capacity to begin something new by virtue of birth is the principle upon which Arendt separates the sphere of equality that marks the *bios politikos* from *zoe,* where embodied differences of all kinds necessarily hold sway.

This observation is well known and has been made by many readers of Arendt.

Here, however, I want to draw attention to an earlier reference in *The Origins of Totalitarianism* to Augustine's understanding of natality, one that Arendt herself never fully develops but that points to another dimension of the event of natality. This other dimension insists on the affirmation of all that she seems subsequently to dismiss from the political space. It calls into question Arendt's strict distinction between the public and the private, between the *bios politikos* and *zoe*. Further, it suggests that the very plurality that Arendt understands as the *conditio sine qua non* of political life is infused with an ineradicable difference or alienness that is inextricably part and parcel of the right to have rights. Her earlier reference to Augustine indicates that *zoe* (unqualified life) must be included in the *bios politikos* and that to exclude it is to commit an originary act of violence that is antithetical to the very existence of the public space and the right to have rights.

GIVENNESS AND HUMAN RIGHTS

Arendt's reference to Augustine occurs in Part Two of *The Origins of Totalitarianism,* at the conclusion of her analysis of imperialism, which ends with an examination of the decline of the nation-state and human rights. In the very last pages of this analysis, Arendt describes those stateless refugees who, having lost their political status as citizens, have lost any and all recourse to human rights. What she and other refugees found was that in the very situation where the declaration of general human rights ought to have provided remedy, just the opposite occurred: "If a human being loses his political status, he should, according to the implications of the inborn and inalienable rights of man, come under exactly the situation for which the declarations of such general rights provided. Actually the opposite is the case" (*OT,* 300). It is in the context of the loss of human rights that Arendt refers to Augustine. I again quote the text:

> The human being who has lost his place in a community, his political status in the struggle of his time, and the legal personality which makes his actions and part of his destiny a consistent whole, is left with those qualities which usually can become articulate only in the sphere of private life and must remain unqualified, mere existence in all matters of public concern. This mere existence, that is, all that which is mysteriously given us by birth and which includes the shape of our bodies and the talents of our minds, can be adequately dealt with only by the unpredictable hazards of friendship and sympathy, or by the great and incalculable grace of love, which says with Augustine, "*Volo ut sis* (I want you to be)," without being able to give any particular reason for such supreme and unsurpassable affirmation. (301)

Here Arendt points to another dimension of the event of natality that seems to be at odds with her concluding reference to Augustine in *Origins*. Rather than emphasizing natality as the capacity for action and for beginning something new, Arendt points approvingly to the Augustinian insight that the event of natality is also about that which is given—indeed, mysteriously given—and which cannot be changed. While she argues that this sphere is usually articulated only in the private sphere, it is important to note that Arendt does not dismiss givenness from the concerns of the public space. Instead, she argues that it must remain "unqualified, mere existence in all matters of all public concern." She insists that givenness should continue to appear, without qualification, in the public space.

In three subsequent places, Arendt elaborates on what she understands by the "given." First, in *The Human Condition*, immediately after discussing the plurality that is the *conditio per quam* of all political life, she refers to the book of Genesis: "But in its elementary form, the human condition of action is implicit in Genesis ('Male and female created He *them*'), if we understand that this story of man's creation is distinguished in principle from the one according to which God originally created Man (*adam*), 'him' and not 'them', so that the multitude of human beings comes be the rest of multiplication" (*HC*, 8). Arendt underscores the point in a footnote to this passage:

> Thus it is highly characteristic of the difference between the teaching of Jesus of Nazareth and of Paul that Jesus, discussing the relationship between man and wife, refer to Genesis 1:27, "Have ye not read that he which made *them* at the beginning made them male and female" (*Matt.* 19.4), whereas Paul on a similar occasion insists that the woman was created "of the man" and hence "for the man" even though he then somewhat attenuates the dependence: "Neither is the man without the woman, neither the woman without the man." (8n)

Pointing out that Jesus links faith to action while Paul links faith to salvation, Arendt suggests that at the very heart of plurality, which she considers to be the essential characteristic of the political, is the givenness of difference and that this difference has everything to do with natality.

Second, at the conclusion of her important essay "Philosophy and Politics," in which she argues that the origin of political philosophy lies in accepting in "speechless wonder the miracle of the universe, of man and of being," Arendt again quotes Genesis 1:27, "the miracle that God did not create Man, but 'Male and female created He them.' They would have to accept something more than the resignation of human weakness, the fact that 'it is not good for man to be alone.'"[1] Here again, Arendt calls for the *political* acceptance of the "miracle of givenness," arguing that the acceptance of this difference is not cause for resignation but is the condition for the very possibility of the human capacity for action.

Finally, in a letter to Gershom Scholem in which she responds to his charge that she has a "cold heart" toward her own Jewishness, Arendt writes: "The truth is I have never pretended to be anything else or to be in any way other than I am, and I have never felt tempted in that direction. It would have been like saying that I was a man and not a woman—that is to say, kind of insane. . . . There is such a thing as a basic gratitude for everything that is as it is; for what has been *given* and was not, could not be, *made*; for things that are *physei* and not *nomoi* (*JP,* 296)."[2] For Arendt, embodiment, including differences in gender as well as differences in ethnicity, like being Jewish, are included in the "birth of the given." These are *physei,* not *nomoi.* To deny givenness would be a form of madness. Arendt further suggests that givenness is at the very heart of human plurality and is the condition for human action. Givenness carries the ethical demand of unconditional affirmation and gratitude—*Amo: Volo ut sis.*

Reflecting on the birth of the given, Arendt argues that from its inception, the Western political tradition has had a profound distrust of this aspect of existence, all too quickly relegating it to the private sphere:

> Since the Greeks, we have known that highly developed political life breeds a deep-rooted suspicion of this private sphere, a deep resentment against the disturbing miracle contained in the fact that each of us is made as he is— single, unique, unchangeable. This whole sphere of the merely given, relegated to private life in civilized society, is a permanent threat to the public sphere, because the public sphere is as consistently based on the law of equality as the private sphere is based on the law of universal difference and differentiation. (*OT,* 301)

Arendt's suggestion that the "given" is relegated to the private sphere not because of its diminutive or privative status in comparison to the reality granted by the light of the public space (as she argues in *The Human Condition*) but because of a long-standing and deep-seated Western resentment toward the singular and the unique is striking. The given, understood as that which is "single, unique, unchangeable," is viewed as a permanent threat to the public sphere, based as it is on the law of equality. Miraculous, hence ineffable, it becomes the alien background of political life: "The dark background of mere givenness, the background formed by our unchangeable and unique nature, breaks into the political scene as the alien which in its all too obvious difference reminds us of the limitations of human activity—which are identical with the limitations of human equality." The alien,

Arendt goes on to argue, "is the frightening symbol of the fact of difference as such, of individuality as such, and indicates those realms in which man cannot change and cannot act and in which, therefore, he has a distinct tendency to destroy" (302).

Whereas in *The Human Condition* Arendt argues that it is the political space with its lawful borders that holds violence at bay, in the above passage, she is arguing that the political space itself has a tendency to destroy that which it cannot change or act upon: "The more highly developed a civilization, the more accomplished the world it has produced, the more at home men feel within the human artifice—the more they will resent everything they have not produced, everything that is merely and mysteriously given them (301)." Let us not forget that these remarks on the political destruction of the alien come at the very conclusion of a long reflection on European imperialism, in which Arendt describes how the Enlightenment ideal of a universal humanity failed completely when European imperialists met with horror Africans, whom they viewed as alien and other than themselves. This leads Arendt to conclude that the "scramble for Africa" reflected the dark and destructive heart of European politics itself that was subsequently unleashed on Europe itself and those "alien" peoples in its midst. And let us also not forget that these remarks on the tendency of Western politics to destroy the "disturbing miracle of the given" (the alien) come at the very end of her analysis of the decline of the nation-state and the rights of man, suggesting that her earlier reference to the "subterranean stream of Western history [that] has finally come to the surface and usurped the dignity of our tradition" might very well be the stream of Western political thought that could not recognize the dignity of that which is other, alien, different. Arendt suggests that it is the disregard of the givenness of human existence that fuels imperialism and, paradoxically, leads to the demise of the modern political space, including its notion of human rights. Finally, let us not forget that Arendt arrives at the "disturbing miracle of the given" in her status as a refugee, an "unqualified" alien who is left with only her mere existence as a human being and who, as a consequence, must flee for her life.

This is all to say that Arendt is well aware of an originary violence at the very heart of the Western political space, an exclusionary violence that dismisses the given, the realm of unqualified mere existence, from the boundaries of the political. As Arendt's analysis of Conrad's *Heart of Darkness* suggests, this original violence haunts the Western political space as it continues to do violence in the name of the political on that unqualified mere existence that is denied entry at its borders. Arendt's analysis of the racialization of the modern political space suggests that this originary violence so haunts the Western political space, is so consubstantial with Western politics, that the modern nation-state constituted through the sovereign homogeneous power of a "people" reduces some of its members to unqualified mere existences. It does so in order to expel them not simply from state borders but from the face of the earth altogether. Genocide, she predicts in 1963, will continue to be the dominant political violence of the contemporary world.[3]

It is strange that Arendt does not develop these earlier reflections on the event of natality and the birth of the given, with its demand of affirming the birth of this "disturbing miracle": *Amo: Volo ut sis.* She chooses to emphasize natality as the capacity to begin and to enact something new. This emphasis on natality as beginning informs her analysis of the private and the public in *The Human Condition,* in which she *seems* to repeat the Western resentment and dismissal of the given. Arendt equates the given with *zoe,* mere life, which she then relegates to the realm of the private, the realm of need and necessity; she also seems to remove it from the political space with its emphasis on freedom, action, and the miracle of new beginnings. It is as if Arendt herself fell blindly into the very subterranean stream of Western history that in her preface to the first edition of the *Origins* she had warned her readers about.

Matters are more complicated, however. For Arendt actually has two different and at times competing notions of the given in her work. The first sense is the one mentioned above, the "disturbing miracle of the given." Arendt first develops this sense in her doctoral dissertation, published as *Love and Saint Augustine;* continues to think it in *The Origins of Totalitarianism;* and returns to it in both volumes of *The Life of the Mind.* In these works, givenness, mere unqualified existence (*zoe*), is understood as the "givenness of appearance itself" and as such must be gratuitously—and gratefully—affirmed. It is this sense that Arendt has in mind when she argues in *The Origins of Totalitarianism* that *zoe* as givenness has been resented, violated, and ultimately excluded from the Western political space. The second and much-better-known sense is found in *The Human Condition* and *On Revolution.* Rather than being relegated to the private sphere by the Western political space, *zoe* becomes the private space that is itself ipso facto identified with *natural* life and the necessities of embodiment, which are more appropriately addressed by the household, or *oikos.* Arendt's seeming agreement with the Greeks that violence may be appropriate to *zoe* in the private domain is startling: "Because all human beings are subject to necessity, they are entitled to violence towards others; violence is the prepolitical act of liberating oneself from the necessity of life for the freedom of world" (*HC,* 31). Here Arendt seems to reduce the given, mere life, to the natural and thus seemingly joins the ranks of those who advocate violence toward it.

I want to suggest, however, that we should be careful about allowing Arendt's second sense of *zoe* and the political consequences she draws from it to occupy the final or even dominant place in her thinking. Instead, what we must observe is her *continual* preoccupation with the first sense of life (*zoe*) as the "disturbing miracle of the given." This allows for a very different understanding of the second sense of *zoe*, albeit one that Arendt herself never explicitly develops. Arendt's analysis of *zoe* and the *bios politikos* in *The Human Condition*, and especially her critique of the social as replacing both the political and private, can be differently understood if grasped on the basis of Arendt's early insistence on the unqualified affirmation of givenness and her recognition that the Western political space is characterized not by affirmation of but by violence toward this dimension of natality. If we follow Arendt's reading of Augustine's understanding of natality, beginning with her doctoral dissertation, we find that she is continually preoccupied with the double miracle of the event of natality, both the miracle of the given and the miracle of beginning. This has great importance for Arendt's formulation of the right to have rights insofar as this double *principium* at the very heart of the right to have rights means that Arendt is rethinking the exclusion of givenness, the expulsion of unqualified mere existence from the political space. More positively, she is insisting that this fundamental exclusion be overcome such that givenness, unqualified mere existence, occupies a rightful place in the political domain. This will in turn have enormous significance in rethinking her account of *zoe* in both *The Human Condition* and *On Revolution*.

ARENDT AND AUGUSTINE:
THE BIRTH OF THE GIVEN

It is important in this reflection on the "birth of the given" to avoid the conclusion that for Arendt the self is given as a fixed or unchangeable datum. Arendt is not invoking a notion of the natural or of a human nature understood as substance, or *ousia*, when she defines the "given" as *physei*. In rejecting a notion of a fixed nature or substance, she is very much indebted to Augustine, insofar as Augustine's understanding of both the divine and the human being marks a radical departure from the Aristotelian notion of substance.

Augustine's *On the Trinity* provides the most complete discussion of his understanding of substance. Augustine begins this work by arguing that in the case of the Highest Being, to be and to exist are the same: "Indeed, in Him [God] it is not one thing to be, another thing to live (as if He could exist without living); nor is it one thing to live and another to understand (as if He could live and not understand); nor finally is it one thing to be happy (as if he could understand and not be happy). Instead, for Him, to understand, to be happy—this is for Him to be."[4] For the Highest Being, essence is coextensive with complete actuality. The problem facing Augustine, however, is how this complete substance can be both one and many. How to account for the trinity? If God is pure substance with no accidents, how can the Son be one with God and yet be created by God? If the Son is created, does this not mean that He cannot be the same as the Father? And if He is not the same as the Father, then are we not faced with two substances, thereby destroying the entire concept of the Trinity? Augustine attempts to answer the dilemma by appealing to Aristotle's categories of relations and substance, reiterating Aristotle's basic argument that substance is that which cannot change and answers the question of "what" the being is.

But Augustine differs from Aristotle in the understanding of relation. As is well known, Aristotle includes relation under the category of accidents. But, Augustine asks, if there is nothing accidental in the highest substance, how do we explain the *relation* between the Father and Son? Augustine states: "There is no question here of an accident, because the one is always the Father and the other is always the Son."[5] The immutability of God necessitates the absence of accidents. Thus, any relation belonging to the substance of the highest being must itself be substantial and immutable.

This raises yet another difficulty concerning the relation of the Trinity. If the relation is itself substantial, then are we still not speaking of two substances, Father and Son, both immutable and unchangeable, and thus denying the uniqueness of the Highest Being? Augustine argues that the relation of the Father, Son, and Spirit is one of attributes that belong not to each person of the Trinity singly but to the Trinity as whole. Thus, there is one substance and three persons in a relation that is not accidental: "Therefore, although the Father is not the Son, and the Son is not the Father, or although the former is unbegotten but the latter is begotten, they do not on that account cease to be one essence, since the relationship between is only made known by these names."[6] At this point, we can see that Augustine has significantly changed the Aristotelian category of both relation and substance; he has posited a relationship that is not accidental, not subject to change, and yet is not simply substantial—that is, inherent in itself. He argues for a notion of substance whose relation is part of its essence.

Augustine's understanding of a relation that is not accidental to the substance extends to his consideration of the human being. Because the human being is an image of the divine being, it also is composed of a trinity: "But just as there are two things, the mind and its love, when it loves itself, so there are two things, the mind and its knowledge when it knows itself. Therefore, the mind itself, its love, and its knowledge are a kind of trinity."[7] The mind, the will, and knowledge are both substantial in themselves yet in a mutual relationship: "These three, therefore, are in a marvelous manner inseparable from one another, and yet each of them is a substance, and all together are one substance or essence, while the terms themselves express a mutual relationship."[8] Augustine further shows how human experience is comprised of two fundamental trinities. First, there is the trinity of the outer person, which includes visible objects (bodies in nature), the sense of vision (which belongs to the human body itself), and the attention of the mind (the will): this is the trinity of sensation. There is also the trinity of the inner person, comprised of memory, understanding, and the will: this is the trinity of cognition. Although the human being experiences both trinities as one and as many, nevertheless a fragmentation marks the unity of the finite creature. This fragmentation is due precisely to the created being's living through time; rather than simultaneity, the finite human being only experiences succession. Most important here is Augustine's formulation of the finite being as a substance that includes relation as part of its essential definition. The person, albeit absolute (*in se persona dicitur*) is also essentially a being related to others (*ad alium*), open to others, and defined as a person by this very relativity. This radically breaks with the traditional concept of the person as a substance; it makes a person dependent for his or her very existence on others. Augustine understands relations with others as essential attributes. The very being of the person necessarily implies relatedness with the world and with others in the world.

Arendt is deeply influenced by Augustine's understanding of the person as essentially related to a finite and changing world. She stresses his insight that human beings "are not simply, but only in relation to something else. . . . When life is seen in its mutability as coming into existence and passing away, and hence as neither altogether being nor altogether notbeing, it exists in the mode of relation" (*LSA*, 52–53). Thus, when she argues that the birth of the given is the birth of the unique, singular, and unchangeable, she is not invoking the givenness of the self as an unchanging nature or substance, given once and for all. Moreover, she follows Augustine in arguing that the character of this relation will depend on what the person loves or desires in the world.

> For the lover is never isolated from what he loves; he belongs to it. . . . Hence in *cupiditas* or in *caritas*, we decide about our abode, whether we wish to belong to this world or to the world to come, but the faculty that decides is always the same. Since man is not self-sufficient and therefore always desires something outside himself, the question of who he is can only be resolved by the object of his desire and not, as the Stoics thought, by the suppression of the impulse of desire itself. (18)

The quest for worldliness changes who a person is. This quest transforms him or her into a worldly being: "In *cupiditas,* man has cast the die that makes him perishable. In *caritas,* whose object is eternity, man transforms himself into an eternal, nonperishable being. Man as such, his essence, cannot be defined because he always desires to belong to something outside himself and changes accordingly" (18). Thus, for both Augustine and Arendt there is no fixed human nature, given once and for all; instead, human beings are always transformed by the objects of their desire.

If Arendt, following Augustine, denies a "nature" to the human being, then what does she understand by the "birth of the given" as "*physis,* not *nomos*"? Furthermore, why does the given need to be affirmed politically? Rather than think *physis* as nature, it would be more illuminating of Arendt's thought to understand *physis* in terms of the *arche,* or the origin of human existence. For the *arche* gives human existence its singular, unique, and unchangeable quality and at the same time infuses this existence with an ineradicable and originary givenness that is not subject to *nomos.* In her doctoral dissertation, Arendt claims that for Augustine "the decisive fact determining man as a conscious, remembering being is birth or 'natality', that is, the fact that we have entered the world through birth. The decisive fact determining man as a desiring being was death or mortality, the fact that we shall leave the world in death" (51–52). On the basis of these two experiences, natality and mortality, Augustine distinguishes two different affections that animate the human being.

On the one hand, our mortality is accompanied by fear and inadequacy, which are the springs of desire. Natality, on the other hand, is accompanied by gratitude, which is the source of memory: "Gratitude for life having been given at all is the wellspring of remembrance, for a life is cherished even in memory" (52). Arendt then quotes Augustine, "Now you are miserable and still you do not want to die for no other reason but that you want to live."[9] Our gratitude is for an origin that is the source of all remembrance but which is itself beyond memory. Moreover, for Augustine, our givenness and our gratitude for the given existence is the source or origin of our seeking and our questioning. On this point Arendt stands with Augustine and against Heidegger:

> Since our expectations and desires are prompted by what we remember and guided by a previous knowledge, it is memory and not expectation (for instance, the expectation of death as in Heidegger's approach) that gives unity and wholeness to human existence. In making and holding present both past and future, that is, memory and the expectation derived from it, it is the present in which they coincide that determines human expectation derived from it; it is the present in which they coincide that determines human existence. (56–57)

It is being-toward-birth that allows for whatever unity and wholeness marks human existence; being-toward-birth is accompanied by gratitude for our very givenness. Paradoxically, being-toward-birth does not yield the resolute human being but instead yields one whose being is now called into question: "To sum up: Man initiates the quest for his own being—by asserting 'I have become a question to myself.' This quest for his being arises from his being created and endowed with a memory that tells him that he did not make himself" (57). Gratitude for what has been given not only takes ontological priority over anxiety toward death, it also at the same time points to the "origin" out of which desire, memory, and questioning emerge.

This memory that reminds us of our dependence on something outside of ourselves arises out of finding ourselves in a world; this "finding" marks the givenness of human existence with an ineradicable estrangement that makes the world a desert for human beings. We are not the principle of our own being; at the same time, we remain estranged from this principle. Moreover, being in the world precedes any explicit love of the world; it is what gives rise to desire and love. One's gratitude at finding oneself in the world predates any activity in the world. Arendt suggests that our capacity for *initium*, for beginning, is dependent upon a prior givenness. This priority of gratitude and affirmation is also true of my relation to my neighbor: "It is not love that discloses to me my neighbor's being. What I owe him has been decided beforehand according to an order that love follows but has not established" (42).

Arendt develops Augustine's understanding of the self as a question that arises in the human being's dispersion "in the manifoldness of the world and [is] lost in the unending multiplicity of mundane data" (23). She argues in her doctoral dissertation that Augustine is not a thinker who seeks to overcome this dispersion by retreating into the self-sufficiency or the inner region of the self:

> For the more he withdrew into himself and gathered his self from the dispersion and distraction of the world, the more he "became a question to himself." Hence, it is by no means a simple withdrawal into himself that Augustine opposes to the loss of self in dispersion and distraction, but rather a turning about of the question itself and the discovery that this self is even more impenetrable than the "hidden works of nature." (25)

Arendt points out that when Augustine grasps that God is "the essence of the heart," he discovers a radical alterity that marks the self "at the heart of the self, that which is in me and but is not me." For both Augustine and Arendt, desire and questioning arise out of a double negative that characterizes the *givenness* of human existence in time: "This questioning beyond the world rests on the double negative into which life is placed. And this double negative (the 'not yet' and the 'no more') means exactly the same as 'before' and 'after' in the world" (70). The before or the not-yet is a preexisting relation that cannot be remembered but is the source of all memory: "the nothingness out of which human existence is created. And life must be a constant referring back to its own origin, this 'not yet,' out of which life has positivity and meaning" (71).

This generative origin that cannot be remembered or reclaimed in desire nevertheless gives rise to memory, desire, and meaning. Memory, desire, and meaning are suffused with this generative origin that cannot itself be explicitly recalled or possessed. One might argue that the entire *Confessions* can be read as the memory and desire for an origin that cannot be recalled but that is both praised and yearned for; it is an origin that gives rise to a self that is split, heterogeneous, and always in relation to something other than itself. Ethically and politically, this givenness must be affirmed. For both Augustine and Arendt, it can only be affirmed and praised, though never justified.

At this point, I want to take up the first part of Arendt's response to Gershom Scholem in which she claims that her being a woman is a *physis* that would be insane to deny. I believe that Arendt is following Augustine in pointing to originary doubleness at work within the generative origin itself. Augustine explicitly argues for a double origin that is both divine and filial, the latter understood as both the Son and Adam. At the same time, Augustine's analysis of the Spirit in his discussion of the Trinity points to an integral third aspect that he leaves relatively undeveloped, except to suggest that Spirit is grace or love. In her analysis of Augustine, Arendt concurs with the notion of a double origin of humanity: "When Augustine asks about the origin of the human race, the answer, as distinct from the self-sameness of God, is that the origin lies in the common ancestor of us all. . . . In this second sense, man is seen as belonging to mankind and to this world by generation. . . . The being of man is understood as derived from a twofold source" (112). While Arendt herself does not posit a divine origin to humanity, she does argue that the generative event of natality is double: "Male and female created he *them.*" As we saw above, Arendt rejects the Paulian interpretation of Genesis in which there is first only Adam in favor of a notion of origin that is always already infused with sexual difference. In Arendt's reading, the origin is necessarily double.

If Adam is the universal dimension (*adam:* everyman) of humanity, then Eve is the dimension of the singular and unique. She is the origin of the alien and the foreign intrinsic to each human being in his or her singularity. Including the feminine as the alien aspect integral to the double ori-

gin allows us to better understand Arendt's argument concerning one's relation to the neighbor:

> Love extends to all people in the *civitas Dei,* just as interdependence ex-
> tended equally to all in the *civitas terrena.* This love makes human relations
> definite and explicit. Coming from the thought of one's own danger that is
> experienced in conscience in God's presence, that is, in absolute isolation,
> this love (*diligere invicem*) also thrusts the other person into absolute isola-
> tion. Thus, love does not turn to humankind but to the individual, albeit
> every individual. (111)

Through the feminine (the *principium* of the alien and the singular),
one is able to love the neighbor in his or her singularity. Through Adam,
human beings are united universally in their common humanity. Through
Eve, each is related to the other in his or her "absolute isolation"—namely,
his or her alienness, foreignness, and uniqueness. This alien and singular
other who is my neighbor is in constant danger and in need of continual
affirmation and care—*Amo: Volo ut sis.*

The *Confessions* suggest that it is only through the feminine (more specifically, the maternal) that one is able to rediscover a *singular* relation to
the divine and to the neighbor. The figure of Monica is central to the *Confessions,* suggesting that it is only through the feminine, her anguish *and her
words,* that the self undergoes a sense of perpetual questioning that marks
the abject state of human finitude.[10] Moreover, she does not remain confined to the household; she is constantly setting out on new journeys to
alien and foreign countries. Augustine recounts his final conversation with
his mother in Ostia as they rest before setting out on the voyage to Africa:
"[When] she asked, 'What is left for me to do in this world?', it was clear
that she had no desire to die in her own country."[11] Monica keeps moving—Carthage, Rome, Milan, Ostia. In the *Confessions,* the maternal is not
a figure of rest, serenity, domesticity; she has no fixed or permanent domi-
cile. In the figure of Monica, the feminine origin that accompanies human
existence is marked as alien, foreign, wandering. Significantly, she does not
die in her own bed but in a strange and foreign land, refusing a return to
her homeland even to be buried.

Thus, it is the feminine aspect of the generative origin that renders
each human being a singular, alien, and estranged presence in the midst of
the plurality that marks the condition of worldliness. Or, to state it differently, *in the very midst of plurality* and human action, there remains the
givenness of the alien, the foreign, the unique and different: "Male and female created he *them.*" As Arendt noted more than once yet managed to
overlook in developing her political thought, at the very heart of plurality

there is difference and not merely repetition. It is this sexually embodied difference that is the condition for all action. Furthermore, Arendt's own analysis suggests that *zoe,* life, is not something that can be domesticated and confined to the *oikos* outside the political sphere; instead, *zoe* is the alien, foreign, and unruly element within the political space itself. If Monica is the figure of feminine life (*zoe*), then her wanderings mark this unruly and alien element at the very heart of the public space.

PHYSIS: ON THE GIVENNESS
OF APPEARANCE

Arendt's insight in *The Origins of Totalitarianism* concerning unqualified mere existence and its relation to the political domain is of utmost importance in understanding the role of givenness in the right to have rights. Arendt refers to the given as "those qualities which usually can become articulate only in the sphere of private life and must remain unqualified, mere existence in all matters of public concern" (*OT,* 302). This point bears repeating: Arendt does not exclude the given from the public space; she argues instead that it must remain "unqualified" in all matters of public concern—if by that term she means "merely existent, without reservations or qualifications of any kind." Further, givenness may be qualified in private life without violation, but never in the public domain. The given and, most especially, embodied differences of all kinds must *not* be taken up in politically qualified ways. To do so would be to politicize embodiment in the most dangerous of ways, the racialization of the body being the most prominent abuse in her analyses.[12]

Arendt suggests that givenness understood as mere unqualified existence is precisely what allows for the radical unpredictability of the new. Our capacity for beginning, *initium,* she argues, is different from the *liberum arbitrium,* or free will, "which decided between things equally possible and given to us, as it were, in *statu nascendi* as mere potentialities, whereas a power to begin something really new could not very well be preceded by potentiality, which then would figure as one of the causes of the accomplished act" (*LMW,* 126). Arendt is emphatic on this point. *Initium* is not the actualization of some potentiality: "The Aristotelian understanding of actuality as necessarily growing out of a preceding potentiality would be verifiable only if it were possible to revolve the process back from actuality into potentiality, at least mentally; but this cannot be done; all we can say about the actual is that it obviously was not impossible; we can never prove that it was necessary just because it now turns out to be impossible for us to imagine a state of affairs in which it had not happened" (139). Insofar as Arendt has already insisted that the human being is *initium,* a beginner, she reaches the same conclusion concerning potency and act in the human being: each individual human being is not a potentiality that can be actualized or qualified in one way or another. As *initium,* each one of us is from the beginning an unqualified existence. Each one of us *is,* and all that can be said is that it is impossible for each one of us not to be. At the very heart of *initium* lies the givenness of a stark, unadorned existence.

This insight guides Arendt in her reading of Heidegger's "Anaximander Fragment," which she reads as an extended reflection on *physis* as givenness. In her exegesis of this fragment, she is particularly interested in Heidegger's understanding of *physis* as unconcealment:

> It belongs to the beings that they arrive from and depart into a hidden being. What can hardly have caused but certainly facilitated this reversal is the fact that the Greeks, especially the pre-Socratics, often thought of Being as *physis* (nature), whose original meaning is derived from *phyein* (to grow), that is, to come to light out of darkness. Anaximander, says Heidegger, thought of *genesis* [becoming] and *phthora* [passing away] in terms of *physis,* "as ways of luminous rising and declining." (190)

Physis is *genesis,* an unpredictable appearing. She understands this genesis in terms of a "contingent causality" in which the *aitia,* the constituting powers of appearing, are the unpredictable causes of appearance. She then emphasizes the next sentence, *"as it reveals itself in beings, Being withdraws."* Arendt's emphasis here is on the original abandonment of appearing itself. More precisely, *physis* is abandonment. Arendt is describing how the *aitia* is in an entirely different ontological relation to what appears. Or to put it more radically: there is no ontological relation between the potentializing power of being and its actual appearance. The emergence of *physis* is not a relation of potentiality and actuality. Again following Heidegger, she argues that "the coming and going, appearing and disappearing, of beings always begins with a disclosure that is an *ent-bergen,* the loss of the original shelter (*bergen*) that had been granted by Being" (190). This loss is *not* the suspension of Being, nor is it the abandonment of Being to itself. To think *physis* (coming-to-be of appearance) without any relation to Being in the form of potentiality—not even in the extreme form of *its* abandonment—is to think the letting-be of appearance as such. This is the radical sense of "givenness": there is no relation to Being—Being itself is the relation. Arendt cites Heidegger as follows: "Presumably, Anaximander spoke of *genesis* and *phthora* [generation and decline] . . . [that is,] *genesis estin* (which is the way I should like to read it) and *phthora ginetai,* 'coming-to-be is,' and 'passing-away comes to be'" (190). *Physis* as *genesis estin* (the genesis of existence) is the appearing of appearance itself. This last un-

derscores the point that the event of natality is a self-originating origin insofar as it cannot point to something more primordial, something pre-orig-
inal that would account for its emergence. As the originary event of becoming, each being comes to itself and begins to be starting from itself in
phyein. This is the radical sense of abandonment: the originary event does
not presuppose anything but itself; it does not presuppose anything else
from which it could be derived. Thus, the originary event, as self-origina-
tion, is the act of beginning. There is no constituting potentiality "behind
the appearance." The originary event proceeds from nothingness; it is an
absolute beginning.

Recently, Giorgio Agamben has undertaken to show the problematic
relation of potentiality and actuality, particularly its relation to national
identity and sovereignty: "Assuming my being-such, my manner of being,
is not assuming this or that quality, this or that character, virtue or vice,
wealth or poverty. My qualities and my being-thus are not qualifications of
a substance (of a subject) that remains behind them and that I would truly
be. I am never *this* or *that,* but always *such, thus. Eccum sic* absolutely. Not
possession but limit, not presupposition but exposure."[13] Appearance is
first of all exposure to the world. This echoes Arendt's first sense of the
"public" in *The Human Condition:* "For us, appearance—something that is
seen and heard by others as well as by ourselves—constitutes reality" (*HC,*
50). Arendt understands this appearance as life, as "being among men"
(*inter homines esse*), and distinguishes it from death, that removal from the
world of appearance (51). Only the second sense of the term "public" denotes a "common world." Prior to and distinguished from the publicity of
a "common world," there is singular, unique life appearing as such, albeit
always lived among other human beings. Appearance as such is a *worldly*
appearance, although not yet a common appearance; for Arendt, the latter
is constituted through the common interests, the *inter-esse,* of a plurality of
human beings. The differing sense of the "public" is the difference between
inter homines esse (being among men) and *inter-esse hominess* (being with
others). Only the latter is properly intersubjective and political.

Everything that is is given, unqualifiedly so. With this emphasis made,
Arendt then formulates a second law that is coeval with the law of plurality: "Everything we know has become, has emerged from some previous
darkness into the light of day, and this becoming remains its *law* while it
lasts" (*LMW,* 191, emphasis mine). *Genesis estin* contains its own law; it is
the law of becoming that holds sway over the abandonment of existence to
itself. Following Heidegger on this point, Arendt argues that as finite beings, we are set adrift in the domain of coming-to-be: "In the beginning,
Being discloses itself in being, and the disclosure starts two opposite move-

ments: Being withdraws into itself, and beings are 'set adrift' to constitute the 'realm (in the sense of the prince's realm) of error'" (192). Agamben calls this the "Irreparable," a term that captures very well the sense of errancy and drift at the very heart of existence itself: "The Irreparable is neither an essence nor an existence, neither a substance nor a quality, neither a possibility nor a necessity. It is not properly a modality of being, but it is the being that is always already given in modality, that *is* its modalities. It is not *thus*, but rather it is *its* thus."[14] As we saw above in Arendt's reading of Augustine, human existence is "irreparable" in the sense of being unjustifiable. Cut off and adrift from any sovereign constituting power or foundation, we are not, however, without law. The law of becoming (*physis*) is the law or rule of anarchic givenness. In her reading of the *genesis estin*, Arendt picks up on Heidegger's explicit reference to the prince's realm, the realm of the political, as the realm where this law holds sway. The law of plurality is coequal with the law of givenness, *physis*.

At the end of her reflections on Heidegger's "Anaximander Fragment," Arendt returns to an insight that inaugurates her reflections on appearance in the first volume of *Life of the Mind*: the instinct for self-preservation is not fundamental to our appearance. The law of becoming that holds sway in givenness is not the law of self-preservation with its accompanying desire for persistence; instead, it is the law of givenness with its accompanying gratitude and sense of delight. Here she agrees with Heidegger's and Nietzsche's critique of the sovereign will, which in its desire for self-preservation and persistence breaks with the law of becoming: "The Will as destroyer appears here, too, though not by name; it is the 'craving to persist', 'to hang on', the inordinate appetite men have 'to cling to themselves.' In this way they do more than just err." Arendt insists that to persist is "an insurrection on behalf of sheer endurance." She understands this to mean that "the insurrection is against order (*dike*); it creates the "disorder" (*adikia*) permeating the "realm of errancy" (193). Disorder is the willful rebellion against the order of becoming in favor of some sort of sovereign endurance. *Adikia* is the rebellion against the lawful appearance of givenness in favor of an insurrectionary sovereign will that seeks to impose itself. Order, or the lawfulness of appearance, is restored by our ability to say "Yes" to the *genesis estin*. Arendt argues that there is no higher affirmation than to say, "'*Amo: volo ut sis*,' 'I love you: I want you to be'—and not 'I want to have you' or 'I want to rule you'" (136). At the very heart of *initium*, therefore, is the unconditional affirmation of the given—*Amo: volo ut sis*.[15] This is also Agamben's thought: "In every thing affirm simply the *thus, sic*, beyond good and evil. But *thus* does not simply mean in this or that mode, with those certain properties. 'So be it' means 'let the thus be.' In other words, it means, 'yes.'"[16]

Here again Arendt's analysis proceeds from the original stratification of the event of natality, in which two different *principia* emerge without coincidence: the *principia* of beginning (*initium*) and the *principia* of givenness. Each of these *principia* gives rise to a different relation: the first is a relation to plurality, the second to uniqueness and singularity. The first is a relation to a common world constituted through the *inter-esse* that both holds together and separates the plurality characteristic of a public space. The second is a relation of *disjunction* between the unique appearance of the singular and this plurality. The stratification of the originary event of natality complicates and renders impossible the attempt at establishing an identity out of this plurality. Again, this is why Arendt insists that the "common" be understood as the "sameness of utter diversity." This is due to the alterity at the very heart of plurality: "In man, otherness, which he shares with everything that is, and distinctness, which he shares with everything alive, becomes uniqueness, and human plurality is the paradoxical plurality of unique beings" (*HC*, 176). This "paradox" that marks the originary event of natality means that plurality cannot be reduced to a "numerical multiplicity." It cannot be thinkable as a closed identity; it is irreducible to a totality. The event of natality contains this paradox: the alterity that marks the *principium* of givenness emerges with the *initium* characteristic of plurality. This paradox has its origins in the originary event through which there is an originary, non-coincident, disjunctive relation of these inseparable, yet distinct, *principia*.

Just as Arendt explicitly criticizes sovereignty from the point of view of *initium* with its inherent capacity for power and action, so too she criticizes sovereignty from the principle of *givenness*, with its unconditional demand of the affirmation of the donation of unqualified singularity. While the *initium* calls into question any claim to sovereignty because it must always be said in the plural, the principle of givenness thwarts any claim to sovereignty (particularly state sovereignty) because it is that which is always outside any identity (including national identity) that could argue for hegemony over unqualified existence. Here another sense of the right to have rights emerges: the right of givenness, unqualified mere existence, to appear and to belong to a political space. This is Agamben's point when he argues that unqualified singularity is something that state sovereignty has never been able to recognize: "The state, as Alain Badiou has shown, is not founded on a social bond, of which it would be the expression, but rather on the dissolution, the unbinding it prohibits. For the State, therefore, what is important is never the singularity as such, but only its inclusion in some identity (whatever identity), but the possibility of the *whatever* itself being taken up without an identity is a threat the State cannot come to terms with."[17] It is this singularity without identity that Arendt is thinking when she thinks the "given," that which Western politics has never embraced and welcomed into its midst. This givenness, in other words, has been the principal enemy of the state. Agamben goes on to assert that "a being radically devoid of any representable identity would be absolutely irrelevant to the State. This is what, in our culture, the hypocritical dogma of the sacredness of human life and the vacuous declarations of human rights are meant to hide. *Sacred* here can only mean what the term meant in Roman law: *Sacer* was the one who had been excluded from the human world and who, even though she or he could not be sacrificed, could be killed without committing homicide."[18] Arendt makes the same point in her essay "The Jew as Refugee," pointing out that the Jew must loyally cling to some national existence or other: "If we should start telling the truth that we are nothing but Jews, it would mean that we expose ourselves to the fate of human beings who, unprotected by any specific law or political convention, are nothing but human beings. I can hardly imagine an attitude more dangerous" (*JP,* 65). The danger is "bare murder," murder outside the legal structure, which Arendt distinguishes from "murder" as a crime that must be punished. Speaking of those who are nothing but naked human beings, Arendt asserts: "Today, the truth has come home: there is no protection in heaven or earth against bare murder, and a man can be driven at any moment from the retreats and broad places once open to all" (90). The state could kill Jews without been seen as having committing murder because Jews had been reduced to a mere unqualified human being and as such were excluded from the human world altogether (*OT,* 442).

This last explains Arendt's elliptical passages in her dissertation on Augustine's belief (*credere*) in the neighbor and her insistence on loving the isolation of the neighbor. For a thinker usually associated with plurality, this is puzzling, even contradictory, unless we keep in mind her emphasis on the political importance of *genesis estin.* Of affirming the isolation of the neighbor, Arendt writes: "This interdependence shows in the mutual give and take in which people live together. The attitude of individuals toward each other is characterized here by belief (*credere*), as distinguished from all real or potential knowledge. We comprehend all history, that is, all human and temporal acts, by believing—which means by trusting, but never by understanding. This belief in the other is the belief that he will prove himself in our common future. Yet this belief that arises from our mutual interdependence precedes any possible proof" (*LSA,* 101). The other can never prove himself or herself in our common future because no proof or justification can be given. And yet there remains belief in the other, a mutual interdependence that can never be justified. Arendt calls this the "indirect community" of plural yet isolated individuals: "Love extends to all people in the *civitas Dei,* just as interdependence extended equally to all in the *civitas terrena*. This love makes human relations definite and explicit. Coming from the thought of one's own danger that is experienced in conscience in God's presence, that is, in absolute isolation, this love . . . also thrusts the other person into absolute isolation. Thus love does not turn to humankind but to the individual, albeit every individual. In the community of the new society the human race as such is not in danger, but every individual is" (111).

Here Arendt states the thought she will develop in *The Origins of Totalitarianism* and *The Jew as Pariah:* it is not the human race that is in danger, but every individual in it. (To view the humanity as a "race" is for Arendt precisely the problem of a "biopolitics," which seeks to qualify *zoe* in all sorts of ways, especially race.) Every individual as unqualified existent, without representable identity, as isolated existence is in danger in a political state characterized by its inability to recognize the murder of those who are nothing but human beings. These individuals live without recognizable identity; they live in the desert, in isolation. This is each of us in our unqualified givenness that makes each person a stranger: "Every single person needs to be reconciled to a world into which he was born a stranger and in which, to the extent of his distinct uniqueness, he always remains a stranger" (*EU*, 308). The isolation of each individual is the inherent strangeness of each unique individual. Arendt, the thinker of plurality, thus argues: "Love of neighbor leaves the lover himself in absolute isolation and the world remains a desert for man's isolated existence. It is in compliance with the commandant to love one's neighbor that this isolation is realized and not destroyed" (*LSA*, 94). There is something prescient in Arendt's emphasis on this aspect of Augustine's thought just before those events in which the world becomes literally for her a desert and every attempt is made to destroy her isolation as a mere human being. It is as if already in her dissertation on Augustine, completed in 1929, she has an uncanny understanding that this commandment to love the neighbor in his or her isolation, in his or her unqualified uniqueness, was about to be violated in the most unimaginable ways. It is if she understood that the state she was living in was about to undertake the final insurrection against this law of *caritas* to love the neighbor in his or her unqualified existence.[19]

For Arendt, *genesis estin,* this inaccessible origin that carries its law with it, is not wholly anterior to the world. *Genesis estin,* unqualified isolated existence, suffuses the plurality that characterizes the political space. For Arendt, natality and mortality are not *merely* natural events. It is important to recall her argument from *The Human Condition:*

> Nature and the cyclical movement into which she forces living things know neither birth nor death as we understand them. The birth and death of

human beings are not simple occurrences but are related to a world into which single individuals, unexchangeable, and unrepeatable entities, appear and from which they depart. Birth and death presuppose a world which is not in constant movement, but whose durability and relative permanence makes appearance and disappearance possible. (*HC,* 96–97)

Here again Arendt suggests that appearance, the life of the individual, cannot be separated from biological life, although the appearance of the individual is at once its biological life and something more insofar as the embodied appearance of the individual life is part of the world:

The word "life," however, has an altogether different meaning if it is related to the world and meant to designate the time interval between birth and death. Limited by a beginning and an end, that is, by the two supreme events of appearance and disappearance within the world, it follows a strictly linear movement whose very motion nevertheless is driven by the motor of biological life which man shares with other living things and which forever retains the cyclical movement of nature. (97)

Especially important in the above passage is Arendt's linking of "life" to the event of natality. At the very outset of *The Human Condition,* she argues that this event breaks with natural temporality—that is, cyclical temporality. As the givenness of appearance, biological life is always already worldly, although not public in the second sense of belonging to a world where our common interests (*inter-esse*) constitute the public. The miracle of givenness is worldly but resists worldly categorization. The givenness of appearance, life, is simply *inter homines esse.* Arendt develops this point in her discussion of the growth and decay of life which, again, she insists are not merely natural phenomenon: "It is only within the human world that nature's cyclical movement manifests itself as growth and decay. Like birth and death, they, too, are not natural occurrences, properly speaking; they have no place in the unceasing, indefatigable cycle in which the whole household of nature swings perpetually. Only when they enter the man-made world can nature's processes be characterized as growth and decay" (97–98). Arendt critiques Marx's view of labor because he overlooks the inherent worldliness of *zoe;* she argues that he understands the givenness of bodily appearance as *natural,* something physiological whose labor is then view as a natural surplus:

When Marx defined labor as "man's metabolism with nature" in whose process "nature's material [is] adapted by a change of form to the wants of man," so that "labour has incorporated itself with its subject," he indicates clearly that he was "speaking physiologically" and that labor and consumption are but two stages of the ever-recurring cycle of biological life. . . . Marx's consistent naturalism discovered "labor power" as the specifically human mode of the life force which is as capable of creating a "surplus" as nature herself." (98–99, 108)

Arendt's phrase "the unnatural growth of the natural," critically denoting the ways in which the political has increasingly given way to the natural, does not refer to the public appearance of givenness in its unqualified existence; instead, her critique of the unnatural growth of the natural in the public space is a condemnation of the ways in which embodiment is reduced politically to the biological, the better to subjugate it for political ends. Further, it depicts the ways in which the political has used naturalistic doctrines to achieve ends that, in turn, increasingly reduce the political to the natural.

Nowhere is the reduction of the political to the natural more evident than in Arendt's analysis of racism and its horrible dénouement in the death camps. For Arendt, racism is the chief political tool used by imperialism to promote the expansion of its economic and political power. The givenness of appearance, both in Africa and in Europe, was reduced to merely natural biological life. In short, imperialist power is for Arendt biopower, a power that uses racial doctrines to destroy the inherent worldliness of *zoe*. Through imperialist biopower, life is reduced to biological life and subsequently subjugated and excluded from any appearance in the public space. The end result, Arendt argues, is the death camps, those "holes of oblivion" where "living corpses" have been expelled not only from the public space but from the earth altogether (*OT*, 434, 451).

It is from her experience of racism and the death camps that Arendt insists on the political affirmation of the givenness of embodied appearance. She insists that the givenness of appearance is at the very heart of the political space: unique, unchangeable, singular, and ultimately alien to any and all representable identities. For Arendt, there is no relegating it to mere biological life that can then be subjugated as a way to consolidate and expand political power. For Arendt, the resentment and destruction of the appearance of givenness characterizes the political space of the West *from its inception*. Because the right to have rights includes the principle of givenness, understood as the appearance of nonrepresentational and unqualified existence, the right to have rights may now be understood as the right to appear of each embodied singular individual who as such remains inherently alien and foreign, isolated from any and all group identities. Thus, Agamben's insistence that the concept of the refugee be severed from the concept of human rights is misguided and not something that Arendt herself would support despite his claims to the contrary.[20] For Arendt, the appearance of the refugee in its unqualified givenness carries an inherent claim of the right to have rights. The right to have rights is not a sacred right bestowed from beyond; it is a human right that emerges from the event of natality itself. It is, so to speak, a birth right and not a natural right; it demands that unqualified human existence rightfully belong to and be protected by the political space.

Conversely, this right entails that the foreign, alien, and nonrepresentational being is capable of being murdered. The human being has inalienable rights—the rights of the alien, the foreigner—not because of any substantial nature (for Arendt, following Augustine, the self is marked by a fundamental heterogeneity and relatedness to the world) but because of the event of natality itself. Politically, this means that it is impossible to separate the *bios politikos* from *zoe*.

In *The Human Race*, Robert Antelme provides an exemplary account of the "almost biological" that provides the limit against which the totalitarian attempt to reduce human life to the merely biological fails. Describing his year as a political prisoner at Buchenwald and Dachau, contested in his very appearance as a human being, Anthelme writes,

> The calling into question of our quality as men provokes an almost biological claim of belonging to the human race. After that it serves to make us think about the limitations of that race, about its distance from 'nature' and its relation to 'nature'; that is, about a certain solitude that characterizes our race and finally—above all—it brings us to a clear vision of its indivisible oneness.[21]

For Anthelme, at the very boundary of the biological, at the very point where the attempt is made to reduce the human being to its biology, the distance, the "almost" that marks the inseparability of *zoe* and the *bios politicos* reveals itself—the individual human being, even if murdered, appears with "absolute clarity" as belonging to a human world and no other: "We have proof of this here, the most irrefutable proof, since the worst of victims cannot do otherwise than establish that, in its worst exercise, the executioner's power cannot be other than one of the powers that men have, the power of murder. He can kill a man, but he can't change him into something else."[22] At the limitation of human appearance, when Antelme experiences the radical solitude of being nothing more than a human being, nonetheless he still belongs to the human world, a being capable of being murdered and not merely killed.

To be sure, Arendt is not always clear on the relation between these two terms. I submit that had she paid more attention to her first readings of Augustine, her political thought, especially evidenced in *The Human Con-*

dition and *On Revolution,* would have developed along different lines, in which she would have better articulated the relation between *zoe* and the *bio politikos.* She first learns from Augustine that embodied differences of all kinds necessarily appear in the *bios politikos.* Instead of being resented and violently destroyed by the political, these differences, she implicitly argues, must be received with gratitude and affirmed as such politically. Augustine suggests that by virtue of the originary natal event all that is associated with *zoe*—embodiment, singularity, the foreign—is always already worldly and yet is constantly in danger of being eradicated. "Male and female created he *them*"—at the very heart of plurality stands the singular and sexually differentiated alien who, in his or her unqualified isolation, is my neighbor. This is the insight that first attracted Arendt to Augustine: in the event of natality each of us is miraculously and gratuitously given as an inescapably alien, singular, and embodied presence that can never be justified or represented, only affirmed and praised—*Amo: Volo ut sis.*

PHYSIS: THE WEB OF APPEARANCE
AND THE GIVENNESS OF A PEOPLE

It is not surprising that Arendt turned to the figure of Rahel Varnhagen after completing her dissertation on Augustine. Nor is it surprising that in her response to Gershom Scholem she framed her remarks in terms of the givenness of both her gender and her Jewishness. In her dissertation and in her *Habilitationschrift,* Arendt is undertaking a sustained reflection on givenness: *physis.*[23] In the first instance, she grapples with a thinker who is unflinching in his reflection on the event of natality as the originary event of givenness. She discovers at the very center of this event inherent uniqueness and difference (*including* sexual difference) with its accompanying gratitude. In the second instance, she grapples with a woman who at any cost is forever trying to escape her Jewishness only to discover at the end that "one does not escape Jewishness" (*RV,* 216). As Arendt indicates in her reply to Scholem, Jewishness is a "given" (*physis*). But what does this givenness entail? What does it mean to be part of a "people" such that escape is not possible? The very notion of this kind of givenness seems completely at odds with Arendt's insistence on the originary givenness of the singular and unique human being.

With this reflection on the givenness of a people, however, Arendt is taking up another aspect of the event of natality—namely, the web of relationships into which each of us is thrown at birth: "The disclosure of the 'who' through speech, and the setting of a new beginning through action, always fall into an already exiting web where their immediate consequences can be felt. Together they start a new process which eventually emerges as the unique life story of the newcomer, affecting uniquely the life stories of all those with whom he comes into contact" (*HC,* 184).[24] Arendt sees no conflict between the uniqueness of the newcomer and the web that catches him or her at birth. She suggests that the uniqueness and singularity of the newcomer is taken up into the web in such a way that the web of givenness is itself altered. This last again reveals the inseparability of the principle of givenness from the principle of *initium*—the given is accompanied by and is inseparable from the opening to the new. Arendt's reflections on the web of givenness have tremendous importance for rethinking the notion of a people, especially in terms of the relationship of culture to the political space and right to have rights.

At first glance, it appears that Arendt is not completely consistent in her account of a people. At several places in *The Origins of Totalitarianism,* Arendt seems to dismiss entirely the reality of a people, associating this notion with a misguided nationalism. For Arendt, the explicit loss of political status that marks the refugee position is already prepared for by the mystique of national sovereignty that implicitly calls for the purification of the nation by the expulsion of the foreign Jewish element from the "national soul." Thus, Arendt argues that "man had hardly appeared as a completely emancipated, completely isolated being who carried his dignity within himself without reference to some larger encompassing order, when he disappeared again into a member of a people" (*OT,* 291). Arendt is extremely critical of such a disappearance, and her criticism suggests that the very notion of a people is to be met with skepticism, if not outright rejection. By contrast, in her essay "The Jew as Pariah," she asserts that "only within the framework of a people can a man live as a man among men, without exhausting himself. And only when a people lives and functions in concert with other peoples can it contribute to the establishment upon earth of a commonly conditioned and commonly controlled humanity" (*JP,* 90). Again and again in her analysis of anti-Semitism as well as the founding of Israel, Arendt insists on the givenness of a people, specifically the Jewish people, which gives them access to the political space not as human beings in general but as who they are—namely, as Jews. Thus, a fundamental paradox arises in her work: on the one hand, she rejects the notion of a national people, while on the other hand she insists on the political inclusion of the Jewish people *as a people.* This paradox continually reappears in Arendt's early thought as she attempts to think the givenness of a people with concrete political rights without falling into the nationalistic framework of a sovereign people—how to think the givenness of a people without falling into racism.

As we take up this paradox, it is important to remind ourselves of Arendt's disagreement with Sartre at the time of *The Origins of Totalitari-*

anism. Arendt's seminal work begins with an analysis of anti-Semitism, and the preface to that work begins with a critique of Sartre's *Anti-Semite and Jew*. Arendt rejects what she understands as "Sartre's 'existentialist' interpretation of *the* Jew as someone who is regarded and defined as a Jew by others" (*OT*, xv). As Richard Bernstein points out in *Hannah Arendt and the Jewish Question*, Sartre's view has at least two pernicious consequences. First, it feeds into the parvenu's desire that if accepted by the anti-Semite he or she will no longer be Jewish; that is, if Jewishness is created by the anti-Semite, who then no longer regards the parvenu as Jewish, then the parvenu no longer is Jewish. Second, if Jewish identity is dependent on anti-Semitism, then those Jews concerned with preserving Jewish identity will, overtly or covertly, be complicit in preserving anti-Semitism.[25] This is Arendt's criticism of the Zionist Herzl, who argued that Jews are constituted as a people by their enemies and thus "our most reliable friends, the anti-Semitic countries, are our allies" (*JP*, 148). Arendt's covert reply is that "anti-Semitism, far from being a mysterious guarantee of the survival of the Jewish people, has been clearly revealed as a threat of its extermination" (*OT*, 8).

Herzl's Zionism is for Arendt yet another version of nationalism that "holds a nation to be an eternal organic body, the product of inevitable natural growth of inherent qualities; and it explains peoples, not in terms of political organizations, but in terms of biological superhuman personalities." Zionism, she argues, is tied up with a tradition of nationalist thinking in which the "sovereignty of the people is perverted into nationalist claims to autarachical existence" (*JP*, 156). Arendt's understanding of a people has nothing to do with inherent organic or biological qualities that identify one group in opposition to another. Her understanding of a people refuses a notion of group identity defined in terms of certain inherent qualities that all members of the group share. It is concerned with what might be called group solidarity, defined in terms of a shared political and historical situation. Most important for Arendt is the requirement that a people understand itself as being in a relationship of solidarity with other political and cultural groups with whom there is a shared concern with justice.

Arendt's biography of Rahel Varnhagen is a kind of *via negative* through which we can begin to understand Arendt's notion of the givenness (*physei*) of a people. Rahel is the figure of someone who tries to escape *as an individual* from her Jewishness. This longing to escape is rooted in a political world in which one had to show that one was an *exception* to one's givenness. Arendt's point is that only extraordinary Jewish *individuals* were allowed some protection from society. Only those Jewish individuals who were exceptional were allowed admittance, "the descendent of an especially exalted people, or else—like Disraeli—[who] sought to validate their people by endowing it with some extraordinary, mystic power" (74). Arendt's critique of the parvenu is precisely that he or she makes it a matter of policy to attempt to penetrate society solely as individuals and this because "every anti-Semite had his personal 'exceptional Jew'" (*JP,* 85). Rahel Varnhagen and her entire generation, according to Arendt, "had already discovered that escape from Judaism was possible only for individuals and that appealing to the spirit of the Enlightenment was no longer of any use" (*RV,* 28). Rahel was able to enter society only as a person and never as a Jew. While she found a limited welcoming of her presence personally, she was always greeted with hostility as a Jew, "for that society had never of its own accorded granted her—as a Jew—the most elementary, most important and minimum concession: equal human rights" (*JP,* 7). Rather than engage in solidarity with other Jews in a political struggle for equal rights, Rahel and others engaged in a personal struggle. Thus, according to Arendt, life for Rahel took on the romantic quality of a great undertaking in which she found herself as an individual engaging with destiny, comparing herself to a great artist for whom "life itself was the assignment" (Preface to *RV,* xvi).

Arendt's analysis of Rahel's life as a life lived as a lie provides us with yet another way to grasp her understanding of the givenness of a people. Her critique of Rahel's individual struggle with destiny is that it is ultimately false; it is a worldless struggle that cannot but then rebound back on itself as a form of introspection rather than a thinking whose proper object remains the world: "If thinking rebounds back upon itself and finds its solitary object within the soul—if, that is, it becomes introspection—it distinctly produces (so long as it remains rational) a semblance of unlimited power by the very act of isolation from the world. . . . Man's autonomy becomes hegemony over all possibilities; reality merely impinges and rebounds" (10). Further, she argues that such introspection engenders mendacity: "Every fact can be undone, can be wiped out by a lie. Lying can obliterate the outside event which introspection has already converted into a purely psychic factor. Lying takes up the heritage of introspection, sums it up, and makes a reality of the freedom that introspection has won. . . . How can a fact mean anything if the person himself refused to corroborate it?" (11). Arendt turns to Mendelssohn, and in this turn we get our first glimpse of her notion of givenness, the givenness of history, a givenness that for Arendt provides the condition for an enduring world: "For to Lessing history is the teacher of mankind and the mature individual recognizes 'historical truths' by virtue of his reason. . . . It is only in Mendelssohn's version that 'historical and rational truths' are separated so finally and completely that the truth-seeking man himself withdraws from history" (12).

Self-deception, she suggests, replaces the givenness of a historical world and one's place in it with an illusionary dream. It attempts to obliterate the given altogether. There is, however, a stubbornness to the given: "Omnipotent as opinion and mendacity are, they have, however, a limit beyond which alteration cannot go; one cannot change one's face; neither thought nor liberty, neither lies nor nausea nor disgust can lift one out of one's own skin" (13). Arendt will return to this issue in her later essay "Truth and Politics," pointing to "stubborn and unwelcome facts" as the truths that "we cannot change; metaphorically, it is the ground on which we stand and the sky that stretches above us" (*BPF*, 264). These stubborn facts are for Arendt always worldly, such that to deny them is to put the world itself in danger: "What is at stake here is this common and factual reality itself, and this is indeed a political problem of the first order" (237). In *The Jew as Pariah*, she first articulates her claim that the lie can destroy the world altogether: "Relationships and conventions, in their general aspects, are as irrevocable as nature. A person probably can defy a single fact by denying it, but not that totality of facts which we call the world" (*JP,* 14). The assimilationist escapes from actual history into an imaginary history of mankind, while the Zionists escape into a doctrine of eternal anti-Semitism. Both escapes, those of the assimilationists and those of the Zionists, avoid taking up the political worldly struggle and the root causes of anti-Semitism.

This, for Arendt, is precisely the problem with Rachel's attempt to escape her Jewishness; it was an escape from her own givenness as a Jew living in a particular historical and political situation. For Arendt, this escape could only be individual: "For the possibilities of being different from what one is are infinite. Once one has negated oneself, however, there are no longer any particular choices. There is only one aim: always, at any given moment, to be different from what one is, never to assert oneself, but with infinite pliancy to become anything else, so long as it is not oneself. It requires an inhuman alertness not to betray oneself, to conceal everything and yet have no definite secret to cling to" (13). Ultimately, Rahel could not live the lie and therefore was not entirely able to embrace the parvenu status. Her insistence on truth-telling left her somewhere between the pariah and the parvenu: "The price demanded of the pariah if he wishes to become a parvenu is always too high and always strikes at those most human elements which alone made up his life. Was it not cause for grief to have no children, no husband her own age, no natural aging and growing gradually weary? What aroused her profoundest indignation was the diabolic dilemma to which her life had been confined: on the one hand she had been deprived of everything by general social conditions, and on the other hand she had been able to purchase a social existence only by sacrificing *nature*" (213, emphasis mine). Again, it is strange that Arendt's later political thought did not include this dimension of political life. A public sphere that guarantees these "natural rights"—the right to one's uniqueness and singularity, the right to embrace one's embodied existence—sexuality, passions, dress, culture—ought to have claimed her attention. Not just general human rights but truly individual rights—the rights of embodied individuals whose sexuality, passions, and culture are recognized in the public sphere—were surely what her own thinking called for.

"Natural right" here must be understood in the context of her discussion of "givenness," in this case, the givenness of an everyday life as a Jew.[26] Arendt claims that the Jews, excluded for centuries from the culture and history of the lands they lived in, had in the eyes of their host peoples remained at a lower stage of civilization. Their social and political situation had been unchanged during these same centuries: everywhere they were only in the rarest and best cases tolerated, but usually they were oppressed and persecuted. Thus, Arendt rejects appeals to grant Jews equal rights on the basis of an abstract humanity. She recognizes that these are not appeals to take up the cause of the oppressed or fellow citizens or even a people with whom anyone felt any ties: "To the keener consciences of men of the Enlightenment it had become intolerable to know that there were among them people without rights. The cause of humanity thus became the cause of the Jews." These were not appeals to allow Jews as Jews a political presence in the public space. Instead, Arendt argues: "the Jews were merely members of an oppressed, uncultured, backward people who must be brought into the fold of humanity. What was wanted was to make human beings of the Jews. Of course it was unfortunate that Jews existed at all; but since they did, there was nothing for it but to make a people of them, that is to say, a people of the Enlightenment" (8).

From this historical background, Arendt asserts the necessity of a different emancipation theory: one must be emancipated as a Jew and as a human being. She points to Schleiermacher and Herder as two of the very few who understood this kind of emancipation. Both favored the naturalization of the Jews as *Jewish citizens,* and both insisted that "there could be no thought of ending the Jewish question." Emancipation for both Schleiermacher and Herder was a political question, that "of incorporating a different nation into the German people and into Europe" (29). But Schleiermacher and Herder were the rare exceptions for Arendt. More in keeping with the sentiments of the time were the views of H. E. G. Paulus, a liberal Protestant theologian and contemporary of Humboldt: "Paulus protested against the idea of emancipating the Jews as a group. Instead, he urged that individuals be granted the rights of man according to their personal merits" (106–107). The destruction of European Jewry began when some Jews, believing they were exceptional, accepted exceptional privileges

from the nation-state, mistakenly thinking that such privileges could protect their human rights. Thus, Arendt asserts, "the terrible and bloody annihilation of individual Jews was preceded by the bloodless destruction of
the Jewish people" (109).

For Arendt, human beings ought to have a right to a nominal rather
than an exceptional existence. Her reading of Kafka's *The Castle* illuminates her concern with the political acceptance of the givenness of the everyday life of a group of people.[27] The hero of this text, she argues, is "plainly
a Jew," not because of any typical Jewish qualities or traits but because the
hero is "involved in situations and perplexities distinctive of Jewish life"
(84). The hero K. seeks to be indistinguishable in order to seek assimilation
with the people of the village. He is refused. This is for Arendt the dilemma
of the would-be assimilationist Jew, "to belong to the people (which is to
belong to those who rule them) or to renounce their protection and seek his
fortune with the masses" (84). K. chooses the latter, "which is to speak for
the average small-time Jew who really wants no more than the rights as a
human being: home, work, family, and citizenship" (85). Significantly,
Arendt, who in *The Human Condition* appears to strictly divide the private
from the public, argues that K. rightly turns to universal human rights "as
the minimum prerequisite of human existence." Among these minimum
prerequisites, Arendt includes the right to work, the right to be useful, and
the right to found a home and become a member of society. She argues that
one ought not to have to become indistinguishable—assimilated—in order
to be granted these rights. The commonplace ought to be rightfully granted
to all and not granted as a privilege to the exceptional Jew who otherwise
occupies the "abnormal" position in society. K. dies exhausted. Arendt is in
complete agreement with Kafka—unless one lives in the framework of a
people, such exhaustion is inevitable. Kafka's affiliation with Zionism lies in
this last point. Arendt points out that he was not a nationalist but "the last
of Europe's great poets" whose work sought to give expression to a human
existence, to being "a normal member of the human society. . . . Men of
goodwill should not be forced to be saints or madmen" (89).

Arendt's criticism of Zionism gives us another clue to the political importance that givenness, in this case the historical and political givenness of
a people, has for thinking the right to have rights. In her perhaps most important essay on the state of Israel, "Zionism Revisited," she is in league
with Kafka, calling for a Jewish homeland rather than a national state.
What is the difference between a "Jewish homeland" and a Jewish state?
The answer, I think, can be found in Orlando Patterson's concept of *natal
alienation*. Describing the condition of slavery in his seminal work *Slavery
and Social Death*, Patterson argues that the slave's natal alienation was the
state of being "culturally isolated from the social heritage of his ancestors.

He had a past, to be sure. But a past is not a heritage. . . . Slaves differed from other human beings in that they were not allowed freely to integrate the experience of their ancestors into their lives, to inform their understanding of social reality with the inherited meaning of their natural forebears, or to anchor the living present in any conscious community of memory." He goes on to argue that when these excommunicated persons did struggle to "reach back for the past, as they reached out for the related living," that meant struggling "with the dominant community, its laws, its heritage, its culture."[28]

While the status of Jews in Europe was far different from that of the state of slavery, nevertheless I submit that there are profound similarities in terms of the cultural alienation Patterson describes, most especially the ways in which natal alienation constitutes what he calls a "social death," which he describes as the lacking of belonging *in his own right* to any legitimate social order."[29] At the very least, Patterson claims, this is the experience of a "secular excommunication." This last certainly describes the situation of both the parvenu and the parish. In the first instance, there is a denial of one's Jewish identity in order to gain social acceptance; in the second, the refusal to deny one's Jewishness leads to excommunication from the social and cultural world. In both cases, neither the parvenu nor the pariah has symbolic access *as a Jew* to the dominant culture.

Still further, if Patterson is right that cultural alienation is in large part constitutive of social death, the state of a living death in society, then it is easier to understand why it was relatively easy for the Jews in Europe to experience so quickly the several deaths that Arendt describes in *Origins of Totalitarianism*: the death of the juridical person, followed by the death of the moral person, followed by the death of spontaneity, and finally, the living dead in those "'holes of oblivion," the death camps themselves. All of these deaths, Patterson's analysis suggests, were made possible by a natal alienation that excommunicated the European Jewry, reducing their existence to a "living death in society."

A Jewish homeland is for Arendt the remedy for overcoming the condition of natal alienation. Her insistence that Hebrew University be at the center of such a homeland indicates that she is thinking of a place where Jewish culture and tradition could flourish and be reclaimed as a living present. And this would be something far different than a Jewish state in which only Jews are full citizens. The latter simply repeats the natal alienation, except that this time Jews are the dominant and alienating force over the Israeli Arabs in their midst.

Throughout her writings on Palestine and Jewish-Arab conflict, Arendt consistently distinguishes between a Hebrew nation and a Jewish people. She rejects the former, arguing that it reflects the fact that the Zionist

movement "was fathered by two typical nineteenth-century European political ideologies—socialism and nationalism" (*JP,* 136). The social revolutionary movement abdicated politically, leaving the course free for what
Arendt calls political Zionism, which shares with all other nineteenth-century political movements an embrace of nationalism: "It never was more
evil or more fiercely defended than since it became apparent that this once
great and revolutionary principle of the national organization of peoples
could no longer either guarantee true sovereignty of the people within or
establish a just relationship among different peoples beyond the national
borders" (141). Clearly Arendt is not arguing for a "national givenness"
when arguing for the givenness of Jewish people. For her the framework of
a national state is such that there are only two alternatives for the solution
of nationality conflicts: complete assimilation or emigration. Arendt
equates national sovereignty with a homogeneous national will in which diversity of ethnicity and culture is not tolerated. By contrast, she insists that
the givenness of a people be understood in relation to others with whom
they share a political and cultural space, and this is something that nationalism refuses to do. Zionism is a nationalism because it too refuses such
recognition of a shared political and culture space with a plurality of heterogeneous peoples: "With an eye only for 'the unique character' of Jewish
history, insisting on the unparalleled nature of Jewish political conditions
which were held to be unrelated to any other factors in European history
and politics, the Zionists had ideologically placed the center of the Jewish
people's existence outside the pale of European peoples and outside the
destiny of the European continent" (155). This kind of thinking she argues, denies the Jewish people its place in European culture, engaging in "a
crazy isolationism gone to the extreme of escape from Europe altogether"
(156). At the same time, Zionism refused solidarity with its Arab neighbors. Arendt suggests that a new opportunity presented itself for Zionism
to consider its "geographical, historical, cultural, and political roots in Europe and at the same to integrate the Jewish people into the pattern of Asi-
atic politics [in an] alliance with the national-revolutionary peoples of Asia
and participation in their struggle against imperialism" (156).[30] As is well
known, this opportunity was rejected and, with chilling clarity about the
Zionist nationalist venture in Palestine, Arendt predicts that "the Arabs
[will] turn against the Jews as the Slovaks turned against the Czechs in
Czechoslovakia, and the Croats against the Serbs in Yugoslavia" (161).

Arendt understands the givenness of a people not in terms of inherent
psychological or biological qualities that would constitute identity but instead in terms of the solidarity of historical, cultural, and geographical
givenness. She discovers in the Dreyfus affair that it was the complete lack
of this solidarity that led to the catastrophe, asserting that "we don't feel en-

titled to Jewish solidarity; we cannot realize that we by ourselves are not so much concerned as the whole Jewish people" (60). Further, she insists that the solidarity of a people must be extended to "the oppressed everywhere even though the historical conditions might be otherwise" (152). By contrast, nationalism understands a people only within the *closed* framework of one's own people and history. In the Dreyfus affair, Clemenceau emerged as the great hero because he understood that "by infringing on the rights of one you infringe on the rights of all" (*OT,* 106). The national and international solidarity of heterogeneous peoples is Arendt's answer to both nationalism and imperialism.

When she refers to her Jewishness as *physis,* Arendt understands it to be historical and cultural givenness. Such givenness, she argues, has the right to equal access to the public space; it has the right to equal participation and protection in the political space. The givenness of a people for Arendt lies in a certain spatiality of condition, a certain historicality of condition, and a certain embodied condition that does not mean determinateness but does indicate diversity and difference that must be politically recognized. This is the case because the archaic event of natality carries the principle of givenness, which is as inherent in the given right of a people, with its unique and diverse culture and history, to appear in the political space as it is inherent in the given right of the unique and singular individual to appear publicly. Again and again in her analysis of anti-Semitism and the Jewish question, Arendt insists on the givenness of a people's right to gain access to the political space, not as human beings in general nor as individuals but as who they are: Jews who are able to gain access to the political space rather than simply assimilating themselves to various social spaces.

Finally, Arendt does not lose sight of the gratitude that accompanies such givenness, arguing that at the end of her life Rahel discovers a longing to be grateful to her givenness as a Jew. Speaking of the parvenu, she writes: "These traits [gratitude and passionate comprehension of the givenness of human dignity]—and Rahel calls them her twin 'unspeakable faults'—the parvenu must discard. He dare not be grateful because he owes everything to his own powers; he must not be considerate to others because he must esteem himself a kind of superman of efficiency, an especially good and strong and intelligent specimen of humanity, a model for his poor pariah brethren to follow" (*RV,* 214).

It is important to remind ourselves again that the principle of givenness is inseparable from the principle of *initium.* Both principles together constitute the right to have rights. Significantly, this means that Arendt is not insisting on rigid boundaries for given cultural and historical traditions. The givenness of a people has nothing to do with the "authenticity" of culture, the latter emphasizing a kind of stasis of cultural and historical traditions that are in need of preservation. Arendt understands the givenness of a people in terms of a web of appearance that is itself always altered by the *initium* of the newcomers who "fall into an already existing web where their immediate consequences are felt" (*HC,* 184).[31] This insight animates Bhaba's insistence that cultural traditions are always engaged with an encounter with newness that is not part of a continuum of past and present—the new is "an insurgent act of cultural translation."[32] This insight also is at the basis of Benhabib's claim that traditions cannot be rooted in the natal community.[33] The principle of givenness is inseparable from the principle of *initium.* The event of natality does not give rise to a "natal community." There are instead singular, alien, and unique newcomers who are caught in differentiated and diverse webs of givenness that are themselves always altered by the appearance of the new. Gratitude for what is given (for the singularity of being, for the otherness of living things, for the webs of appearance, for the unfolding of time, for the earth itself) is accompanied by the joy of inhabiting together with a plurality of others a world where the unpredictability of the new remains an ever-present possibility. Both gratitude and joy animate the right to have rights.

FOUR

The Predicament of
Common Responsibility

It is precisely because the tyrant has no desire to excel
and lacks all passion for distinction that he finds it
so pleasant to rise above the company of all men;
conversely, it is the desire to excel which makes men
love the world and enjoy the company of their
peers, and drives them into the public business.

Hannah Arendt, *On Revolution*

Natality's archaic principle is double: the principle of beginning and the principle of givenness. Consequently, the animating affection is itself double: pleasure in the company of others and gratitude for givenness. Arendt especially celebrates pleasure, understanding it to be the animating affection of public life. In *On Revolution,* she approvingly quotes John Adams: "Wherever men, women, or children, are to be found, whether they be old or young, rich or poor, high or low, wise or foolish, ignorant or learned, every individual is seen to be strongly actuated by a desire to be seen, heard, talked of, approved and respected by the people around him." This desire to be seen and heard by others gives "a feeling of happiness they could acquire nowhere else" (*OR,* 119). Following Adams, Arendt locates the ethical dimension of public life in this desire: "The virtue of this passion he called 'emulation', the 'desire to excel another', and its vice he called 'ambition' because it 'aims at power as a means of distinction.' And, psychologically speaking, these are in fact the chief virtues and vices of political man" (120).

Arendt locates the source of this desire in the archaic event of natality itself: the event of natality carries with it the desire to appear. She goes so far as to call this desire an "innate impulse" as compelling as the fear that accompanies the urge for self-presentation: "It is indeed as though everything that is alive—in addition to the fact that its surface is made for appearance, fit to be seen and meant to appear to others—has an *urge to appear,* to fit itself into the world of appearances by displaying and showing, not its 'inner self" but itself as an individual" (*LMT,* 29). Drawing on the research of Swiss biologist and zoologist Adolph Portmann, Arendt argues that this urge to appear cannot be explained in functional terms. Instead, she suggests, the urge to appear is gratuitous, having to do with the sheer pleasure of self-display. Human beings, who have a concern with an enduring image, transform this urge to self-display into a desire for self-presentation that she argues involves a "promise to the world, to those to whom I appear, to act in accordance with my pleasure" (36). The gratuitousness of this pleasure indicates that appearance itself carries the double affection of gratitude and pleasure. Public happiness, therefore, is the pleasure of appearing in a common world that delivers us from obscurity. It is the pleasure of being visible—being seen and recognized by equals; it is the pleasure of our own image granted only through the perspectives of others. Finally, this pleasure is the animating bond of the "we"; it provides an animating or dynamic basis for the political bond, or what Arendt calls "the solidarity of all peoples" (*OT,* 161). This animating double affection of pleasure and gratitude is for Arendt at the very heart of the right to have rights.

At the same time, Arendt's formulation of human rights emerges out of a reflection on radical evil. Speechless horror, not beauty or pleasure, marks the contemporary experience of wonder.[1] This horrifying wonder (*thaumazein*) at the human capacity for evil animates her entire thought. Paradoxically, a thinking that has its origins in the horror of the twentieth century concludes by insisting on a notion of public happiness. This tension between the horror in the face of what humans are capable of and the human capacity for pleasure in the company of others is for Arendt the predicament of common responsibility and is the subject of this final chapter.

In what follows, I first examine Arendt's analysis of radical evil in *The Origins of Totalitarianism,* arguing that her analysis of "totalitarian hell" is essential to understanding the hellish violence of radical evil. Julia Kristeva's recent work on Arendt (*Hannah Arendt*) is especially important because her concept of abjection illuminates Arendt's claim that the superfluousness of the modern human being accounts for the emergence of radical evil. Kristeva's concept of abjection suggests that the banality of radical evil is the ever-present threat to the fragility of human affairs *precisely because* of the event of natality.[2] Two inseparable moments constitute the event of natality: the abject desolation that carries the relentless threat of radical evil and the activity of beginning that allows for the transformation and fragile redemption of finitude itself, a transformation that holds at bay but never eradicates the threat. Arendt's politics of natality emerges out of these two inseparable moments of the event of natality, offering the only possible remedy to the threat of radical evil. That remedy, as we shall see, modifies our relationship to temporality, which, in turn, allows for a transformed sense of the solidarity of humanity through the affective bond of political friendship. Political friendship, for its part, animates the right to have rights.

RADICAL EVIL, ABJECTION, AND THE HORROR OF HUMANITY

For Arendt, a sense of shame is all that seems to remain of any sense of human solidarity. Arendt argues that this sense is the prepolitical or nonpolitical expression of the insight that "in one form or another men must assume responsibility for all crimes committed by human beings and that all nations share the onus of evil committed by all others" (*EU*, 131). The international solidarity of humanity lies in this almost unbearable burden of global responsibility; it prevails only if it faces up to the human capacity for evil. Arendt, however, is not arguing that evil is an inherent trait of human beings. In her 1945 review of *The Devil's Share*, she takes issue with the argument that good and evil are inherent in the human condition, involved in a gnostic fight for dominance. Radical evil does not point to a demonic nature; instead, it is a *capacity*.[3] The problem for Arendt is that the Western tradition has not faced up to humankind's very real capacity for incalculable evil, preferring instead to see evil as a kind of nothingness—a lack of Being or the Good.

Although initially speechless, Arendt does attempt to articulate her wonder at the horror of the twentieth century; she names it *hell*. The terror and total domination of the death camps is the fabrication of hell on earth: "Three kinds of concentration camps can be very aptly divided into three types corresponding to three basic Western conceptions of a life after death: Hades, Purgatory, and Hell." Hades, Arendt argues, corresponds to "those relatively mild forms, once popular even in non-totalitarian countries, for getting undesirable elements of all sorts—refugees, stateless persons, the asocial and the unemployed—out of the way." She goes on, "Purgatory is represented by the Soviet Union's labor camps, where neglect is combined with chaotic forced labor. Hell in the most literal sense was embodied by those types of camps perfected by the Nazis, in which the whole of life was thoroughly and systematically organized with a view to the greatest possible torment" (*OT*, 445).

Arendt suggests that the emergence of total domination and terror is the hubristic appropriation of religious limits, specifically the belief in hell.[4] Totalitarian domination materializes this belief by incarnating it in immanence:

Suddenly it becomes evident that things which for thousands of years the human imagination had banished to a realm beyond human competence can be manufactured right here on earth, that Hell and Purgatory, and even a shadow of their perpetual duration, can be established by the most modern methods of destruction. . . . Nothing perhaps distinguishes modern masses as radically from those of previous centuries as the loss of faith in a Last Judgment; the worst have lost their fear and the best have lost their hope. Unable as yet to live without fear and hope, these masses are attracted by every effort which seems to promise a man-made fabrication of the Paradise they had longed for and of the Hell they had feared. The one thing that cannot be reproduced is what made the traditional conceptions of Hell tolerable to man: the Last Judgment, the idea of an absolute standard of justice combined with the infinite possibility of grace. (446–447)

In this passage Arendt points to the symbolic function images of heaven and hell have played in political thought since Plato's *Republic:* they arouse both our longings and our fears. Religion, however, puts heaven and hell beyond the reach of human fabrication. Arendt suggests that although the modern political space is marked by an abyss opened by the loss of its theological underpinnings and a loss of belief in the Last Judgment, these representations still continue to play a political role at the level of our hopes and fears.[5] The separation of the theological from the political opens the way for the secularization of these representations, bringing them down to—and up to—earth: "The totalitarian hell of totalitarianism proves only that the power of man is greater than they would ever have dared to think, and that man can realize hellish fantasies" (446). In her essay "Religion and Politics," Arendt reiterates this insight: "In totalitarian states we see the almost deliberate attempt to build, in concentration camps and torture cellars, a kind of earthly hell" (*EU,* 383).

In a 1951 letter to Karl Jaspers, Arendt clarifies this point, suggesting that the totalitarian vision of hell is an attempt to establish an omnipotent presence on the earth itself:

What radical evil is I don't know, but it seems to me it somehow has to do with the following phenomenon: making human beings as human beings superfluous. . . . This happens as soon as all unpredictability—which, in human beings, is the equivalent of spontaneity—is eliminated. And all this in turn arises from—or, better, goes along with—the delusion of the omnipotence (not simply the lust for power) of an individual man. If an individual man qua man were omnipotent, then there is in fact no reason why men in the plural should exist at all—just as in monotheism it is only god's omnipotence that makes him ONE.[6]

Arendt calls this desire for omnipotence the "madness for the superlative,"[7] a madness that brings God down to earth in the figure of a particular omnipotent individual. Arendt is clear in her letter to Jaspers that this madness is very different from the desire for power that is found, for example, in Hobbes. For Hobbes, she argues, the desire for power remains comparative, relative to the power of other human beings. By contrast, the desire for omnipotence is a rejection of plurality altogether in favor of "being one," a godlike power on earth that desires absolute rule.

The hell of radical evil lies in refusing symbolic transcendence, represented by religious and moral limits, substituting instead the fantasies of immanent ideologies and omnipotent dreams. Here we grasp the full import of Arendt's insistence that radical evil requires a move from "everything is permitted" to "everything is possible." "Everything is permitted" faces the death of God but still recognizes the exigency of judgment, of making a distinction between the permissible and the impermissible, even if the impermissible is emptied of any absolute measure. "Everything is possible" refuses both the death of God and the exigency of judgment. It reestablishes an omnipotent presence on earth without any hope of pardon or grace.

The rage against the symbolic, the collapse of transcendence into immanence, is also true of totalitarianism's relation to the law. Arendt insists that these regimes are not lawless. Totalitarian regimes, she argues, "claim to obey strictly and unequivocally those laws of Nature and History from which all positive laws always have been supposed to spring" (*OT*, 461). Raging against the constraining yet absent symbolic law, totalitarian politics "promises justice on earth because it claims to make mankind itself the embodiment of the law" (462). Totalitarianism substitutes another law that would be incarnate and reassuring because the law can now be known—it literally dwells among us, having been brought down to earth.

This becomes evident in the trial of Eichmann. Arendt reports that Eichmann suddenly declared that he had lived his whole life according to the Kantian moral imperative. At first Arendt is affronted at such an outrage against Kant. Upon further examination, however, Arendt grasps that what Eichmann actually did was to pervert the Kantian law, substituting the will of Hitler for the universal and transcendent law of reason:

> [Eichmann] had not simply dismissed the Kantian formula as no longer applicable, he had distorted it to read: Act as if the principle of your actions were the same as that of the legislator or of the law the land—or, in Hans Frank's formulation of the "categorical imperative in the Third Reich," which Eichmann might have known: "Act in such a way that the Führer, if he knew your action, would approve it. (*EJ*, 136)

In Eichmann, Arendt is confronted with the specificity of the general claim she first made in *Origins in Totalitarianism*: the terror of radical evil and total domination is possible through the perversion of the symbolic dimension of the law—that is, a human being becomes its embodiment, its sovereign will: "In Kant's philosophy, that source [of the law] was practical reason; in Eichmann's household use of him, it was the will of the Fuhrer" (137).

The perversion of the law is accompanied by a perversion of desire. While attention has been paid to Arendt's analysis of the role of duty for the law-abiding citizen, it is not often noticed that her analysis of the dutiful citizen concludes with a discussion of the inseparability of Eichmann's sense of duty from his resistance to the *temptation* to do good: "Evil in the Third Reich had lost the quality by which most people recognize it—the quality of temptation. Many Germans and many Nazis, probably an overwhelming majority of them, must have been tempted *not* to murder, *not* to rob, *not* to let their neighbors go off to their doom. . . . But, God knows, they had learned how to resist temptation" (150). The resistance to desire occurs through the fascist imperative of obedience and sacrifice. It is an imperative delivered most forcefully by what Eichmann terms the "winged words" of Himmler, who was the most gifted, Arendt argues, at solving the problem of conscience—the desire to resist evil. The effect of these winged words on Eichmann was one of an elation in which the slogans and watchwords were no longer felt to be issued from above but instead seemed to be self-fabricated: "And you could see what an extraordinary sense of 'elation' it gave to the speaker the moment it popped out of his mouth." Arendt points out that whenever the judges "tried to appeal to his conscience, they were met with 'elation', and they were outraged as well as disconcerted when they learned that the accused had at his disposal a different elating cliché for each period of his life and each of his activities" (53).

Eichmann's voice of conscience was not silenced—it was carried away, caught up in the voice of another; his voice had literally been "voiced over" with the voice of Himmler. Not only does Eichmann's elated voice of conscience identify the law with the will of Hitler but also, and at the same time, his desires and fantasies become identified with Hitler's. The elated voice of conscience told Eichmann to ignore his own desire and dutifully carry out the law of the land: "And just as the law in civilized countries assumes that the voice of conscience tells everybody 'Thou shalt not kill,' even though man's natural desires and inclinations may at times be murderous, so the law of Hitler's land demanded that the voice of conscience tell everybody: 'Thou shalt kill', although the organizers of the massacres knew full well that murder is against the normal desires and inclinations of most people" (150). Citing the court's judgment, Arendt points out that for justice to be based on the voice of conscience, orders to be disobeyed must be "manifestly unlawful" and unlawfulness must "fly like a black flag above them as warning: 'Prohibited!'" She goes on to argue, however, that in Hitler's regime

> this "black flag" with its "warning sign" flies as "manifestly" above what normally is a lawful order—for instance, not to kill innocent people just because they happen to be Jews—as it flies above a criminal order under normal circumstances. To fall back on an unequivocal voice of conscience—or in the even vaguer language of the jurist, on a "general sentiment of humanity" (Oppenheim-Lauterpacht in *International Law*, 1952)—not only begs the question, it signifies a deliberate refusal to take notice of the central moral, legal, and political phenomena of our century. (148)

These phenomena of our century are at least two: the fragile status of the law and its subject. The transformation of the transcendent law into perverse immanence attests to the fragility of the law in modernity. Eichmann's all-too-easily-replaced voice of conscience, an elated voice in which he identifies with both the law and the desires of the Führer, points to the fragile identity of the modern subject.

Eichmann's sacrifice of personal desire through the elated voice of conscience is accomplished, Arendt argues, by turning basic instincts such as the instinct of pity, whereby we recoil at the suffering of others, back upon the self: "The trick used by Himmler . . . consisted in turning these instincts around, as it were, in directing them toward the self. So that instead of saying: What horrible things I did to people! the murderers would be able to say: What horrible things I had to watch in the pursuance of my duties, how heavily the task weighed upon my shoulders!" (106). Himmler's "trick," accomplished through slogans and stock phrases (e.g., "My honor is my loyalty"), is effective because it promises the unity of the subject if only the subject releases his or her desires. In sacrificing desire for duty, the subject has the fantasy of a stable and fixed identity. In a perverse departure from Rousseau, self-pity allows for a unified *amour propre* only on the condition that the subject becomes an elated and at the same time a dutiful instrument of the other's fantasies.[8]

Scant attention has been paid to how the fragile identity of the modern subject informs Arendt's analysis of radical evil, which she understands as the attempt to eliminate spontaneity from the human race. It is the attempt to reshape human nature itself by doing away with the very unpredictability that lies at the root of human freedom and action, the attempt to stabilize human behavior in order to allow the law of history or the law of nature to progress unhindered. Arendt notes: "The camps are meant not only to exterminate people and degrade human beings, but also [to] serve the ghastly experiment of eliminating . . . spontaneity itself as an expression of human behavior and of transforming the human personality into a mere thing, into something that even animals are not; for Pavlov's dog, which, as we know, was trained to eat not when it was hungry but when a bell rang, was a perverted animal" (*OT,* 438). The terror of totalitarianism is thus involved in the inseparable perversion of both the law and human subjectivity. It perverts humanity by eliminating the capacity for action, the capacity for new beginnings, and it perverts the very meaning of the law, transforming it from its traditional sense as that which provides limits and boundaries to human action into that which is limitless and constantly on the move. The movement of the law now requires that human beings be static and fixed. Arendt locates the appeal of totalitarian ideology in its claim to carry out the law of nature or history—that is, in the longing for a fixed and stable identity:

> Just as fear and the impotence from which fear springs are antipolitical principles and throw men into a situation contrary to political action, so loneliness and the logical-ideological deducing the worst that comes from it represent an anti-political solution and harbor a principle destructive of all human living-together. . . . The 'ice-cold reasoning' and the mighty tentacle of dialectics which 'seize you as in a vise' appear like a last support in a world where nobody is reliable and nothing can be relied upon. It is the inner coercion whose only content is the strident avoidance of contradictions that seems to confirm a man's identity outside all relationships with others. (478)

Fascist ideology promises a readymade, unified identity—fixed, static, without contradiction, and utterly reliable. The madness for the superlative is mirrored in the desire of the individual desolate human being who also wants to reject the plurality (the two-in-one) at the very heart of the self in favor of a completeness, an integrity promised in submitting to a fantasy of omnipotence. For Arendt, the appeal of this promise of unity has its roots in the modern phenomenon of superfluousness. Radical evil, she writes in *Origins,* "has appeared in connection with a system in which all men have become superfluous in some way." It is the desolation of individuals who are economically superfluous and socially uprooted that provides the conditions for radical evil. A peculiar kind of loneliness is key to understanding this evil: "Loneliness, the common ground for terror, the essence of totalitarian government, and for ideology and logicality, the preparation of its executions and victims, is closely connected with uprootedness and superfluousness which have been the curse of modern masses. . . . To be uprooted means to have no place in the world, recognized and guaranteed by others, to be superfluous means not to belong to the world at all" (475). While Arendt argues that superfluousness is a peculiarly modern phenomena, she affirms that "we have only to remind ourselves that one day we shall have to leave this common world which will go on as before and for whose continuity we are superfluous in order to realize loneliness, the experience of being abandoned by everything and everybody" (476).

A radical superfluousness or abandonment marks human finitude itself. For Arendt, following Heidegger, we are set adrift in the domain of coming-to-be. Giorgio Agamben calls this the "Irreparable," a term that captures very well the sense the abandonment at the very heart of existence itself.[9] As Jean-Luc Nancy points out, the word banality comes from the same root as the word abandon: *bannum.*[10] Banality is the condition of a humanity that has been forsaken, banished—we are holes of oblivion. In the past, this desolation or banality has been covered over by the tripartite structure of authority, tradition, and religion. Modernity, Arendt argues, is marked by the splintering of this structure in which our desolation appears at the very center of our existence. This oblivion is the secular ordeal of modernity. The banality of radical evil is the refusal to endure this ordeal.

Here we arrive full circle, back to Arendt's claims that the idea of humanity is terrifying. Arendt argues against the popular notion that the more we know about each other, the more we will come to like each other. On the contrary, Arendt writes, "The more peoples know about one another, the less they want to recognize other peoples as their equals, the more they recoil from the ideal of humanity" (235). Arendt's insistence that the element that most unites us—humanity—is also the element that causes terror and a certain recoiling is important. The ideal of humanity, purged of all sentimentality, demands that human beings assume political responsibility for all crimes and evils committed by human beings. At the same time, she argues, this demand is terrifying; this is "the predicament of common responsibility" (236). Our predicament lies in the double face of humanity: our Janus-like humanity is at once what unites us in common responsibility *and* what causes us to recoil in terror. The recoil, Arendt suggests, is because of our banality, our desolation. Our terror lies in facing up to our lack of being, our being holes of oblivion. In a letter to Gershom Scholem, Arendt argues that radical evil spreads like a fungus on the surface of human existence (*JP,* 128). The horror of the banality of radical evil is precisely this fungus-like quality that attempts to fill in the cracks and holes of human finitude with dreams and deliriums of fabricating the absolute on earth. It necessarily lies on the surface because it attempts to cover over the abyssal nature of human existence.[11] Critical of the Western tradition's understanding of evil as nothing, a lack of the good, Arendt suggests that the banality of radical evil lies in the disavowal of our own nothingness, our own desolation and impossibility of being. Of utmost importance here is Arendt's insight that the event of natality itself carries with it this desolation and the ever-present threat of radical evil as the refusal of this desolation.

ARENDT AND KRISTEVA: NATALITY AND THE PREDICAMENT OF COMMON RESPONSIBILITY

Arendt's discussion at the conclusion of Part Two of *The Origins of Totalitarianism* of the Western resentment of the given, a resentment rooted in the fear of the frightening fact of difference, answers the problem she raises at the outset of her analysis in Part One of anti-Semitism: "There is hardly an aspect of contemporary history more irritating and mystifying than the fact that of all the great unsolved problems in our century, it should have been this seemingly small and unimportant Jewish problem that had the dubious honor of setting the whole infernal machine in motion" (*OT,* 3). *The Origins of Totalitarianism* can be read as a reflection on the Western fear of the alien that crystallizes in the terror of totalitarianism. Arendt critiques the demand that individuals and groups assimilate because it only covers over a simmering resentment toward the alien and different and thus is dangerous. In the context of assimilation, she shows how Jewishness is first embraced socially as "exotic" before being made a vice to be expunged. Certainly her analysis of the scramble for Africa, specifically the terror she sees as implicit in the Enlightenment ideal of humanity with its concomitant demand for common responsibility, has its basis in the Western fear of what remains unique, singular, and foreign.

It is strange that Arendt does not develop her early reflections on Augustine's insight that the event of natality is not only *initium* but also givenness. She chooses instead to emphasize natality as the event of beginnings. This emphasis informs her notion of public happiness, which is rooted in our appearance among others with whom we have the capacity to enact something new. Arendt's emphasis on freedom and action fails to consider her early argument that political space, with its capacity for the new, has *as its condition* givenness, with its imperative of mysterious gratitude. In her later writings, Arendt splits the doubleness of the originary event: the miracle of beginnings infuses the speech and action of the public space, while the disturbing miracle of the given and the alien is relegated to the barbaric and violent space outside the city walls. Thus, the ethical demand for gratitude succumbs to violence outside the law. Paradoxically,

Arendt's reflections on imperialism suggest that this violence toward given-ness returns to haunt the very heart of the Western political space. Pleasure in the company of others gives way to Hobbesian grief.

Kristeva points out that for Arendt "social philosemitism always ended by adding to political anti-Semitism, that mysterious fanaticism without which anti-Semitism could hardly have become the best slogan for organiz-ing the masses. That leaves us to explore the psychological components of this 'fanaticism.'"[12] Arendt is aware of the dangers of social "philosemitism." She clearly understands that the flip side of assimilation is the expulsion of what resists the demand that it belong. The embrace of the exotic alien al-ready contains the seed of its annihilation. Arendt suggests that this violent expulsion of the alien lies in the failure to acknowledge the imperative of gratitude toward givenness. It is precisely this failure that preoccupies Kris-teva. She argues that it can grasped only be as "psychoanalytic anthropol-ogy": "If we resist the traditional safeguard of religions, with their focus on admonishment, guilt, and consolation, how can our individual and collec-tive desires avoid the trap of melancholic destruction, manic fanaticism, or tyrannical paranoias? The absence of a psychoanalytic anthropology is clear. . . . Arendt curiously appears not to see why it matters."[13]

Kristeva is therefore extremely critical of Arendt's formulation of plea-sure as the *sole* animating affection of the political bond. More specifically, she questions Arendt's almost total disregard of the psychological vice that Arendt herself admits could animate the political space—ambition that uses power to dominate others. Why domination rather than emulation as a means to achieve distinction? Why *not* domination, with its accompany-ing fear and violence, rather than emulation with its pleasure in appearing in the company of others? Kristeva argues that Arendt delves into the Christian concept of authority, particularly the fear of hell that underlies it, and considers the fearful interplay between rewards and punishments that is the substratum of faith to be "the only political element in traditional re-ligion," and yet she concerns herself with neither the psychological foun-dation of this dynamic nor the indispensable support that it offers the po-litical bond. Kristeva asks: "Are perhaps all political bonds based on an arousing fear?"[14] With this question, Kristeva seems much closer to Hobbes than to Arendt. Fear, not pleasure, is the animating affection of the public space, allowing us to better understand the violence that seems endemic to it. If pleasure is possible, it is more hard won than Arendt admits.

Kristeva locates the source of their disagreement in Arendt's refusal to consider the event of natality as an *embodied* event. Her refusal to consider embodiment, Kristeva argues, leaves Arendt's analysis unable to consider the sadomasochistic desire at the very heart of natality and consequently as something that must be considered in our formulation of the right to have rights: "We should remember, however, that the refusal to contemplate the uniqueness of the body and the psyche is what drove Arendt to refuse to acknowledge the role played by sado-masochism in the experience of violence."[15] Kristeva argues that while Arendt locates the cause of modern violence in the decline of the political, "which engenders coercion to compensate for its weakness and to gain strength," nevertheless "the psychological element—sadomasochism in particular—would have enriched her analysis with an important element that would help us grasp more effectively the 'conditions' or the 'crystallization' of the phenomenon she describes."[16] Kristeva suggests that while Arendt develops the psychological virtue of excellence at the foundation of our pleasure in the company of others, she ignores completely its vice—our desire to dominate and inspire fear in those to whom we bring nothing but grief.

Kristeva criticizes Arendt for ignoring the uniqueness of the body and the psyche and for being therefore unable to give an adequate account of domination, with its accompanying fear and violence. Kristeva suggests that Arendt's refusal to acknowledge the role played by sadomasochism, the violence beneath our desires, renders her incapable of accounting for our refusal of the imperative of gratitude toward givenness, with its accompanying pleasure in appearing with others, that is Arendt's prerequisite for action. In the preface to her three-volume work on feminine genius, Kristeva argues that she needs to discuss Melanie Klein after her analysis of Arendt precisely because Klein's account of natality offers a more complete understanding of both sadomasochism and gratitude. In *Tales of Love*, Kristeva notes that Klein, "the bold theoretician of the death drive[,] is also a theoretician of gratitude seen as 'an important offshoot of the capacity for love' necessary for the acknowledgment of what is 'good' in others and in oneself."[17] Kristeva's analysis of Klein shows how despite the violence underneath our desires gratitude for the given is possible—how it is possible for us, even while acknowledging the violence of the sadomasochistic drive, to find pleasure in the company of others. This analysis has tremendous significance for grasping the animating affection at work in the right to have rights.

Further, Kristeva's psychoanalytic anthropology leads again to the archaic event of natality, and she understands this event through her notion of abjection. In *Powers of Horror*, Kristeva argues that abjection is the "the result of a primary natality, the birth pangs of a body becoming separated from another body in order to be."[18] This primary natality "preserves what existed in the archaism of a pre-objectal relationship, in the immemorial violence with which a body becomes separated from another body in order to be—maintaining that night in which the outline of the signified thing vanishes and where only the imponderable affect is carried out."[19] Kristeva claims that abjection rises from a primal repression when the infant struggles to separate from the mother's body that nourishes and comforts, from the ambivalent struggle to establish a separate bodily schema while still seeking a unity with the mother's body, which it seeks to incorporate. Thus, the subject enters into language from a background of conflict between attraction and repulsion with regard to an image of the pre-oedipal archaic mother. Our desolation, our banality, is due to the very first birth pangs of embodiment, the traces of which we carry with us into linguistic natality. Prior to linguistic natality, the subject is located in processes that cannot be named. The identity/nonidentity of the subject as a signifying process exists prior to birth into the symbolic order of language under the father's law.

Through her reading of Klein, Kristeva rethinks the "pre-objectal" nature of immemorial violence. On the basis of her reading of Klein, she argues that immemorial violence (the sadomasochistic or death drive) that is part of primary natality is always already in a relation to an object: "[Klein] was more receptive than were other analysts to the hypothesis of a death drive in the baby that responded to its fear of being destroyed. . . . And yet, by considering the drive to be more psychological than biological, Klein added that the death drive manifests itself only through its relation to an object."[20] Contrary to Freud (and this has tremendous import for Kristeva), Klein argues that the drives are not directionless aimless psychic energy; from the beginning they bear the stain of language, "a rudimentary presence of symbolization at the level of drives is at work."[21] For Klein, because the drives are already endowed with rudimentary symbolization, fantasy is borne by sensation as well as affect. This claim, Kristeva points out, distances Klein from Lacan: "The Kleinian phantasy includes elements that her followers would seek to conceptualize. Lacan, for his part, adopted a decidedly Greek approach by shifting psychic representation toward the appearance and toward the visibility of the *eidos*. Psychoanalysis today focuses on this clinical and conceptual exploration of the transverbal archaic realm that Melanie brought to our attention, a realm that belies the ideal of visual representation."[22] While for Lacan, especially the Lacan of the mirror phase, fantasy is a project of the idea into the appearance of the drive, Klein claims that fantasy is "saturated with the reality of drives and with such primary 'contained' contents as greed and envy."[23] While Lacan gives primacy to the signifier, Klein argues for incarnation—a sensation that is both an affect and a representation.

Significantly, this is Kristeva's first step away from Hobbes. While Hobbes argues that language "bears the stain of the passions,"[24] thereby necessitating that the declaration establishing the principle that the sovereign be secured by the sword, Kristeva argues that the passions always already bear the stain of language, thereby necessitating further supplementation of the negative (symbolization) rather than force. Hobbes suggests that there is a primordial violence to the passions that forever eludes language, thereby requiring that language be forever supplemented by the sword. Kristeva, by contrast, argues that the primordial violence of the passions is always already mediated by a rudimentary symbolization, thereby offering the possibility that further symbolization (the work of the negative) may make the sword unnecessary.[25]

Furthermore, while for Freud the object is the "object of an instinctual aim,"[26] Klein argues that the object is always something more; it is an object-relation involving the fantasies and anxieties of the infant. Thus, from the beginning, drives are always directed toward others—whether real or imaginary. Further, and again contrary to Freud, Klein argues that the body is not the source of drives. Drives do not originate as tensions within the body that then affect the mind, whose basic function is to meet the needs of the body by eliminating drive-tensions and preserving a state of equilibrium. Instead, Klein argues, the body is the means of the expression of drives. This insight is important to Kristeva, who points out that "Klein claimed the anamorphosis of the body into the mind, of sensations and affects into signs and vice versa." Kristeva argues that Klein posited that "flesh-and-soul are forever linked in the heart of the human being" and that in so doing, she "revived flesh within the word, and she privileged the body of drives and passions within the imagery."[27]

Kristeva's embrace of the Kleinian understanding of drives against both Freud and Lacan greatly clarifies her notion of abjection, which, she argues, must be understood at the level of the "translinguistic primary level of the drives."[28] Drives are already object-seeking and imbued with rudimentary symbolization at the level of bodily process; they already have a relationship to reality. Neither the ego nor the id nor signifiers are needed to give the semiotic drives coherence or direction. From the start, the infant's psychic universe is consumed with a "primary symbolization" in an affective reality. Insofar as all drives seek objects, there is no pre-objectal relationship. Kristeva argues that "from the moment of birth, the drive engages in a binary expression: sensation/affect and the object *coexist,* and the presentation of the object clings to sensation."[29] Through her reading of Klein, Kristeva also rethinks "immemorial violence" as the sadistic fantasies of the archaic ego "directed against the inside of the mother's body constituting the first and basic relation the outside world and reality."[30] Rather than understanding abjection as the border conflict between the semiotic drives and symbolic processes, Kristeva's reading of Klein relocates the border conflict of abjection in the conflict between the inherent destructiveness of the sadistic aim and the reparative aim of gratitude.

Kristeva follows Klein in arguing that the death drive in the infant is a response to the fear of being destroyed. Kristeva understands the death drive as an innate destructive impulse that is the natural response to frustration. The infant experiences frustration as total annihilation and projects that feeling toward the object; it is a paranoid destructiveness because the infant feels as if the attack is directed at him or her. Thus, the infant splits aggression off from love and experiences it as paranoia. The infant's fear of death is fear of disintegration in the face of its own hatred. This persecutory anxiety is the infant's fear of its own aggressive impulses arising from the death drive, which is in conflict with the life drive. There is a redirecting of aggression toward an external object, initially the mother's breast: against its own anxiety, the infant projects the death instinct outward. The sadism of the archaic ego prolongs original anxiety, which is the infant's "fear for its life" and which has a strong oral desire to devour what threatens it: "What is manifested at the very beginning of life, returns to the subject with the same content but with a different target: it is not I who wishes to devour, for I am afraid of being poisoned by the bad breast in which I projected my bad teeth—such is the logic of the sadistic fantasy that corresponds to primary paranoid-schizoid anxiety."[31] Thus, sadism defends itself from its own destructiveness and envy through splitting the object into a good breast and a bad breast. As Jessica Benjamin points out in her analysis of the sadomasochistic drive, "aggression ends up doing *outside* what [it] would otherwise do *inside:* reducing the world, objectifying it, subjugating it."[32]

Sadomasochism is always accompanied by envy, the outward manifestation of the death drive. Envy, a word that derives from the Latin phrase for "to cast an evil eye upon," spoils the good object for someone else or deprives them of it. Here Kristeva points to Augustine, for whom envy was the worst sin because it "opposes life itself."[33] Envy attacks and destroys pleasure in the self and others and is the direct expression of the destructive impulses specifically directed against the source of life. The good object, usually first the mother's breast, is resented and hated because it is endowed with life-giving qualities that the infant depends upon for its survival. But the good object is not always available. When this lack occurs, the infant attacks and spoils the good object, which is now blamed for the deprivation. Envy is the destruction of what is other, foreign, and strange; it is the hatred of what one cannot have and what is unobtainable.[34]

The subject that emerges from this unnamable point of division is split, identifying its previous, fragmentary experience that only exists as affect—bare want, loss that is unrepresentable—with the mother's body. Before desire—the movement outward from a self to the objects toward which it is directed—there are drives that involve pre-oedipal semiotic functions and energy discharges that connect and orient the body to the mother. Abjection is the moment of separation, the border between the "I" and the other, before an "I" is formed; it is want or lack itself—an unassimilable nonunity experienced by one who is neither in the symbolic order nor outside of it. This much of Lacan is preserved in Kristeva.

Abjection is the place between signs; it is a trace, a rhythm, an excess or disturbance that destabilizes and threatens to undermine all signifying processes. Abjection therefore is that place "where the subject is both generated and negated, the place where his unity succumbs before the process of charges and stases that produce him."[35] Thus, the emergent subject is infused with negativity, an alterity that determines its emergent subjectivity. And this negativity is both pleasurable and painful; it is the source of both creation and meaning and of absence, estrangement, desolation. The latter are important; Kristeva stresses that abjection ought not to be "designated as such, that is, as other, as something to be ejected, or separated."[36] Abjection is therefore associated with the disintegration or, perhaps more precisely, the heterogeneity that exists at the very heart of the self.

We must now find our way back to Arendt, for whom our long and complex digression into psychoanalytic anthropology would only have been a painful—yet necessary—experience. Important in Kristeva is the affective dynamic of attraction and repulsion with the mother's body in the labor pains of emerging subjectivity. For Kristeva, abjection is the moment of separation, and it is always double. It is the feeling of loathing and disgust the subject has in encountering certain matter, images, and fantasies— the horrible, to which it can only respond with aversion, with nausea and distraction—and it is at the same time the feeling of fascination, drawing the subject toward it in order to repel it. Kristeva argues that abjection is above all *ambiguity.* While it releases a hold, it does not radically cut off the subject from what threatens it. On the contrary, abjection acknowledges the subject to be in perpetual danger. Kristeva points out that while Arendt is aware of Hitler's fascination with the *Protocols of the Elders of Zion* (it is said that he knew them by heart), she misses the abjection that drives Hitler's interest. Kristeva argues that "Nazi propaganda proceeded by negatively identifying with an enemy slated for death while at the same time imitating him with a hateful fascination."[37] Hitler does not denounce the protocols but seeks instead to establish an exact replica in reality, designating the Jew as his worst enemy in a delirious and yet fascinated revulsion.

Moreover, Kristeva insists that abjection is a historically and culturally specific response to the fragility of the law; it is tied to the secular ordeal of modernity and the collapse of the religious foundation of the political order. She agrees with Arendt's analysis of the immanent status of the law in totalitarian regimes but criticizes her for not taking into account the sadomasochistic dimension that accompanies the fragility of the law and contributes to the fabrication of hell on earth:

> Arendt touches upon the theme of sadomasochism when she delves into the Christian concept of authority, particularly the fear of hell that is its basis. [But she does not] analyze the specific fate of the alchemy between fear and authority that operates at the heart of the secularized modern world, which has clearly left the fear of hell behind but which has in no way diffused the sadomasochistic spirit of what Arendt cautiously refers to as the "frailty of human affairs."[38]

Arendt is not unaware of the instinct for submission, the alchemy between fear and authority, at the heart of the human psyche. In *On Violence*, Arendt observes: "If we were to trust our own experiences in these matters, we should know that the instinct of submission, an ardent desire to obey and be ruled by some strong man, is at least as prominent in human psychology as the will to power, and, politically, perhaps more relevant" (*OV,* 39). Yet Arendt does not explore this instinct, nor does she articulate its political relevance.

Kristeva, on the other hand, suggests that in modernity the political relevance of this desire for submission (what she calls the sadomasochistic dimension) lies in the instability of the symbolic dimension of the law that manifests itself in abjection: the permeability of the inside and the outside boundaries, the weakness of cultural prohibitions, and the crisis of symbolic identity. She specifically links the fragility of the law that is exposed in abjection to a crisis of authority. This crisis manifests itself in phobia, an elaboration of want and aggression: "In phobia, fear and aggressivity . . . come back to me from the outside. The fantasy of incorporation by means of which I attempt to escape fear . . . threatens me nonetheless, for a symbolic, paternal prohibition already dwells in me. . . . In the face of this second threat. . . . I attempt another procedure: I am not the one that devours, I am being devoured by him."[39] To offset the fear associated with the weakness of the symbolic order, the phobic subject regresses to the narcissistic fantasy of fusion with the maternal body. Yet this fantasy is threatening because the subject is always already in the symbolic order governed by the paternal prohibition of incest. Thus, the fantasy is inverted—rather than devouring the mother (the fantasy of incorporation which promises jouissance and the escape from fear), the subject fantasizes that it is being devoured. This phobic fantasy then constructs an imaginary other who becomes the metaphor of the subject's own aggression. Insofar as the phobic fantasy is always culturally and historically specific, fascist regimes are able to mobilize these phobic fantasies and bring them to bear on the social body. Kristeva argues that "the imaginary machinery is transformed into a social institution—and what you get is the infamy of fascism."[40]

Most important, the phobic fantasy operates at the level of drive rather than desire—it is the unleashing of the death drives onto the social body. However, insofar as the subject is already in the symbolic (the paternal prohibition is in effect), these drives *postdate* desire. Kristeva's analysis of phobia allows us to better understand Himmler's successful trick of inverting basic instincts such as the instinct of pity. Reading this inversion through Kristeva's analysis, Himmler's "winged words" produce phobic fantasies at the level of drives incorporated by a subject who has the double fantasy of incorporation and unity (fusion with the maternal/social body) and projection that displaces aggression onto the imaginary other (the Jew) who now seems to threaten from without. Eichmann is caught up in elation—the phobic fantasy that demands only that he sacrifice his desire (the temptation to do good) and carry out his duty to the law. However, this law mobilizes at the level of drives and fantasies rather than at the level of the symbolic. The phobic fantasy is mobilized by "winged words" that hollow out language with an infinity of significations, substituting clichés and slogans that operate at the level of drives—"phobia is a metaphor that has mistaken its place, forsaking language for drive and sight."[41]

In an age where power and the symbolic "exclusionary prohibition" no longer belong to the ultimate Judge—"God[,] who preserves humanity from abjection while setting aside for himself alone the prerogative of violence"—Kristeva argues that the exclusionary prohibition now belongs to discourse itself, which is now the location of the "prohibition that has us speak." The fascist and racist discourses of Louis-Ferdinand Céline and Hitler give legitimacy to hatred as they rage against the monotheistic symbolic law (itself infused with negativity and loss) and substitute in its place another law that would be "absolute, reassuring, and fully incarnated."[42] At the same time, seeking to resecure the boundaries of the immanent law, this discourse turns the Jewish body, which is deliriously viewed by Céline and Hitler as the embodiment of the monotheistic symbolic law, into the rejected site of all forces of negativity, loss, and dissolution.

In her reading of Céline's pamphlets, Kristeva shows how his writings transform an experience of abjection and the fragility of the law into the phallic ambition to name the unnamable. Céline's anti-Semitism and fascism can be seen, therefore, as "a kind of parareligious formation" into a fantasy of "the immanence of substance and meaning, of the natural/racial/familial, of the feminine and the masculine, of life and death—a glorification of the Phallus that does not speak its name."[43] Raging against the symbolic law, Céline substitutes an immanent substantial law in the phantasmatic revitalization of the social body: "Again carrying out a rejection, without redemption, himself forfeited, Céline will become body and tongue, the apogee of that moral, political and stylistic revulsion that brands our time."[44] Like Arendt, Kristeva gives the name hell to the horror of the fascist discourse: "This is the horror of hell without God: if no means of salvation, no optimism, not even a humanistic one, looms on the horizon, then the verdict is in, with no hope of pardon."[45]

Kristeva provides a much-needed supplement to Arendt's understanding of the event of natality that allows us to see the ambiguous and fragile status of this event, out of which arises the predicament of common responsibility. The frailty of human affairs arises first out of the abjection of a primary natality, an abjection that means we must face the ever-present threat of the banality of radical evil, which can be traced to a radical abandonment—a desolation inherent in embodiment itself. Kristeva reminds us that the Arendtian "second birth" (linguistic natality) is not only inseparable from this first birth but also bears within it traces of these primary birth pains. At the same time, it would be a mistake to think the abjection of primary natality as itself inherently delirious or evil. Kristeva agrees with Arendt: evil is a capacity of human beings, not an inherent trait. Evil is our capacity for self-denial, which is fundamentally a denial of abjection. More precisely, the banality of radical evil lies in our inability to live with the abject—to live with the ambiguity, abandonment, and negativity that infuses the event of natality at both its levels, bodily and linguistic.

At the same time, Kristeva points out that the death drive is always in the service of the life drive. For Kristeva, the concern with the life drive, "what is 'full of birth,'" constitutes the proximity between Melanie Klein and Hannah Arendt:

> Though on the opposite side of the spectrum from Hannah Arendt, Melanie nevertheless appears to have shared Arendt's concern for the sort of life that emerges through the revelation and accompaniment of that which threatens it. "Full of birth," as Arendt would put it, which Melanie showed herself to be through the therapeutic relentlessness that pervaded her incisive interpretations—and also through the privileged mode she assigns to the death drive, which is first described as a sadistic desire, as a type of envy, as she would later put it. In sum, the death drive is a condensation of love and hatred, otherwise known as paroxysmal desire.[46]

Kristeva's reference to Arendt at this point in her analysis of Klein is significant. As I maintained at the outset of this chapter, all three thinkers—Arendt, Klein, and Kristeva—are preoccupied with the archaic event of natality, although Arendt splits the doubleness of this *arche* and in doing so does not develop an account of the gratitude toward the given that is inherent in this event. While Arendt recognizes two drives at the heart of human existence—gratitude and anxiety—she does not fully see both as part of the event of natality. Rather, following Augustine she locates fear and anxiety as being toward death, while she sees gratitude as arising out of a primordial memory of the event of natality. Kristeva, following Klein, shows how both anxiety (associated with the death drive) and gratitude (which, like Arendt, Kristeva locates in memory and mourning) are part of the event of natality. For Kristeva, Klein is able to offer a way to repair this splitting, thereby offering an account of how the sadistic and destructive drive can be transformed into the reparative life drive that manifests itself in gratitude toward the given.

While the life drive is always caught up in a struggle with the death drive, Kristeva argues that it manifests itself first in the depressive position in which the infant retains a memory of the good object and acquires nostalgia for it. This memory is comparable to mourning: "But because this love is a love of devouring that is heavily laden with sadistic drives, the feeling of losing the good object is buttressed by a feeling of guilt over having destroyed it by assimilating it."[47] This feeling of mourning or depression mobilizes the desire "to make reparation to objects," by which the infant "imagines that he can undo the nefarious effects of his aggression through [his mother's] love and care for him."[48] This is the "wager of gratitude" that is the indispensable precondition for developing the capacity to make reparation that is "concomitant with the loss of the object in the depressive position."[49] The fragmentation and splitting characteristic of the paranoid-schizoid position is transformed into gratitude for the "whole object," whose ineradicable loss is now mourned. The transformation of the violence of the death drive into gratitude occurs through the mourning of the object, which remains forever foreign and alien in a primordial and irrecoverable separation.[50]

Kristeva's work can be read as offering an extensive argument for the ethical-political imperative of gratitude for the foreign, the alien, and the given in a world that has undergone the death of God. (And could we not read the violence of the twentieth and twenty-first centuries as the destructive fury of the paranoid-schizoid position, which, having experienced the anxiety of this loss, refuses to mourn but instead furiously seeks to destroy what remains unobtainable or in a phobic fantasy constructs an imaginary, fascistic other who becomes the metaphor of its own aggression?) Kristeva argues in *Strangers to Ourselves* that "the issue of foreigners comes up for a people when, having gone through the spirit of religion, it again encounters an ethical concern. . . . The image of the foreigner comes in the place and instead of the death of God."[51] Rather than seeing foreignness as the political facet of a violence that is excluded and that thereby underlies the political space (as Arendt sees it), it becomes a challenge or call for the gratuitous embrace of the alien: "Living with the other, the foreigner, confronts us with the possibility or not of *being an other.* It is not simply—humanistically—a matter of our being able to accept the other, but of *being in his place,* and this means to imagine and make oneself other for oneself."[52]

In her analysis of Klein, Kristeva shows how violence can give way to gratitude for what remains wholly other and alien. Rather than projecting the envy of the death drive outside, gratitude is able to mourn and make reparation to what is always foreign both *outside* and *within.* Contrary to Arendt, therefore, Kristeva shows how the archaic origin is always double but no longer dangerously split. While the human psyche contains an "unfathomable destructiveness," it also has the capacity for "reparation and love with gratitude, as opposed to sadism, the tyranny of the superego and envy."[53] Gratitude, Kristeva argues, is the "quieter side of Thanatos," and she adds: "Anxiety has not disappeared, but it chooses another domain: rather than splitting and fragmenting, rather than destroying and tearing into pieces, anxiety is tolerated as a source of pain relating to the Other and a source of guilt about having taken pleasure in hurting him."[54] Living with the foreign and the alien therefore involves the psychic discomfort of nostalgia and guilt, but it no longer inspires resentment and violence. Pleasure is possible through gratitude and a call for love—"desire always bespeaks anxiety with much intensity; only in quieter times does it become a source of pleasure in which case it is still prepared to seek delight through love and gratitude."[55]

I want to return once again to Kristeva's question to Arendt: "Are perhaps all political bonds based on an arousing fear?" Recall that Kristeva raises this Hobbesian question in the context of what she understands as a secularized modern world that has "clearly left the fear of hell behind but which has in no way diffused the sadomasochistic spirit of what Arendt cautiously refers to as the 'frailty of human affairs.'"[56] Through her reading of Klein, however, we see that Kristeva allows for another possibility of an archaic affection (arising out of the event of natality) that animates the right to have rights and that, in turn, following Montesquieu, would be manifest in the laws and institutions instituted through this right. We have seen that Kristeva defines gratitude as that which diffuses (but does not replace completely) the sadomasochistic spirit of human affairs. Gratitude for the foreign and alien, for what is given as the wholly other and unobtainable, ought to be the animating archaic affection of the political bond in a secularized, postreligious world.

At the same time, Arendt answers Kristeva's question by arguing for a transformation of temporality. Given the human capacity for evil, rooted in our banal denial of abjection, Arendt suggests that the only possible remedy for the modern secular world is to change our relationship to time through a politics of natality.

THE PREDICAMENT OF COMMON RESPONSIBILITY AND A POLITICS OF NATALITY

Arendt begins *The Human Condition* and her analysis of the *vita activa* with a distinction between eternity and immortality. While her discussion of immortality is often read as an argument for heroic deeds and speech that distinguish the actor in the public realm and thereby ensure, through remembrance, his or her endurance in time, close examination of Arendt's argument reveals that she is not so much interested in the endurance of individual deeds as she is in the endurance of humanity itself. Immortality, she argues, is the concern of those beings who are mortal: "Imbedded in a cosmos where everything was immortal, mortality became the hallmark of human existence. Men are 'the mortals'" (*HC*, 18).

Mortality marks the division between life and death; it marks a cut in time. Human beings move "along a rectilinear line in a universe where everything, if it moves at all, moves in a cyclical order" (19). This transformation of cyclical into rectilinear time distinguishes the human being from other animal species. To put it more strongly: to be fully human requires a transformation of time. This transformation, Arendt argues, is accomplished only insofar as mortality is linked to a concern with immortality, the latter being inseparable from a political life: "Without this transcendence into a potential earthly immortality, no politics, strictly speaking, no common world and no public realm is possible. . . . But such a common world can survive the coming and going of the generations only to the extent that it appears in public. It is the publicity of the public realm which can absorb and make shine through the centuries whatever men may want to save from the natural ruin of time." In negative terms, Arendt argues that the decline of modern humanity is inseparable from the decline in a concern with immortality and the public world: "There is perhaps no clearer testimony to the loss of the public realm in the modern age than the almost complete loss of authentic concern with immortality" (55).

Immortality is therefore a political achievement that institutes an enduring common world. Neither a religious sentiment nor a hope founded on the fear of death, the desire for immortality is the desire for a common world that delivers us from eventual obscurity. It is the desire to be visible—to be seen and recognized by equals; it is the desire for our own image granted only through the perspectives of others. Far from celebrating a politics of heroic individualism, Arendt's emphasis on immortality is rooted in the desire to appear—that is, the desire to be: "The term 'public' . . . means, first, that everything that appears in public can be seen and heard by every-body and has the widest possible publicity. For us, appearance—something that is being seen and heard by others as well as by ourselves—constitutes reality" (50). The fulfillment of this desire depends on the existence of a plurality of others who share a common world. Citing Aristotle, she argues that "to men the reality of the world is guaranteed by the presence of oth-ers, by its appearing to all; 'for what appears to all, this we call Being,' and whatever lacks this appearance comes and passes away like a dream" (199). Our very sense of reality "depends utterly upon appearance" in a common world, the reality of which "relies on the simultaneous presence of innu-merable perspectives and aspects in which the common world presents it-self and for which no common measurement or denominator can ever be devised" (57).

Arendt goes so far as to call this desire to appear an innate impulse as compelling as the fear that accompanies the urge for self-preservation: "It is indeed as though everything that is alive—in addition to the fact that its surface is made for appearance, fit to be seen and meant to appear to oth-ers—has an *urge to appear*, to fit itself fit into the world of appearances by displaying and showing, not its 'inner self' but itself as an individual" (*LMT*, 26). As we observed earlier, Arendt argues that this urge to appear cannot be explained in functional terms; instead, she suggests, the urge to appear is gratuitous, having to do with the sheer pleasure of self-display. Human beings, who have a concern with an *enduring* image, transform this urge to self-display into a desire for self-presentation, which, she argues, in-volves a "promise to the world, to those to whom I appear, to act in accor-dance with my pleasure" (*LMT*, 36). The hypocrite, on the other hand, is one who breaks his or her promise to act in accordance with pleasure.[57]

The division between the natural and the mortal/immortal being, therefore, coincides with the first division between the private and public realm. To return to Kristeva's question posed at the outset of this section: for Arendt, the only way our individual and collective desires can avoid the fanaticism and madness of radical evil is through the political institution of a different form of time—the time of immortality—rooted not in re-ligion or fear of death but in the desire for an enduring image and mode of appearance. This is a desire met only in a public space with an irreduc-ible plurality of others with whom we promise our pleasures rather than assert our needs. Kristeva herself suggests that Arendt's highly controver-sial distinction between the social and the political be understood against the background of this transformation of temporality. She asks whether Arendt's distinction between *zoe* and *bios* is not another way to articulate the distinction between needs that link the subject to an archaic realm—that is, its dependence on the mother—and desires that afford the dangerous freedom of bonds with other people in the space of appearance: "To transform the nascent being into a speaking and thinking being, the maternal psyche takes the form of a passageway between *zoe* and *bios,* between physiology and biography, between nature and spirit."[58] She quotes Arendt —"The 'nature' of man is 'human' only to the extent that it gives him the possibility of becoming something highly unnatural, that is, a man" (*OT,* 455)—and argues that Arendt's distinction between *zoe* and *bios* is rooted in her analysis of the death camps where the metamorphosis of human beings into nature served to transform them into living cadavers.[59] As the human being emerges from the event of natality, he or she is a beginner, which means that the nature of the human being is inherently flexible and open-ended. At the same time, this capacity can all too easily be foreclosed. Arendt is fond of quoting Montesquieu: "Man, this flexible being who submits himself in society to the thoughts and impressions of his fellow-men, is equally capable of knowing his own nature when it is shown to him and of losing it to the point where he has no realization that he is robbed of it."[60] The loss of the human occurs whenever the attempt is made to stabilize the "nature" of the human being: to make it unified, complete, without contradiction or heterogeneity. This illuminates Arendt's claim that our very sense of reality is dependent upon a plurality of others with whom we appear. Arendt is not thinking the common world as one established in a reciprocity of identifications. Rather, the commonality of the world emerges out of the irreducible nonintegration of different standpoints—"sameness in utter diversity" (*HC,* 57). At the same time, self-identity is constituted only through these irreducible and innumerable perspectives. Arendt suggests that the endemic duality (the two-in-one) that characterizes the self is made one only in its appearance with others who remain heterogeneous in their appearance: "For this ego only exists in duality. And this ego—the I-am-I—experiences difference in identity precisely when it is not related to the things that appear but only related to itself. (This original duality, incidentally, explains the futility of the fashionable search for identity)" (*LMT,* 187).

Finally, Arendt suggests that the political institution of the temporality of immortality must be accompanied by an affectivity that provides an animating or dynamic basis for the political bond, or what Arendt calls "the solidarity of humanity."[61] Arendt goes much farther than Kristeva in understanding the need for a *political* remedy for the fantasies and deliriums that accompany the banality of radical evil. Kristeva seems still to appeal to fear and authority when thinking the affectivity of the political bond when she asks "Are perhaps all political bonds based on an arousing fear?"[62] She seems still too close to a kind of Hobbesian position wherein the dynamic of fear and authority found in religion is transposed into the fear of the sovereign with the introduction of the modern Leviathan. (It should not go unnoticed that in *Leviathan,* Hobbes's chapter "On Religion" directly precedes the all-important chapter "Natural Condition of Mankind as Concerning their Felicity and Misery.") While Kristeva would certainly not follow Hobbes in the direction of the sovereign—indeed, her work on abjection and on the stranger directly calls Hobbes into question—it does seem that she is not able to think something other than fear as the animating political bond.

Arendt clearly rejects fear as the affect capable of instituting the political bond, arguing that fear is a *nonpolitical* emotion rooted in the isolating self-interest of the individual human being.[63] While Arendt agrees that fear can be used as a political tool for dominating individuals, it cannot be the animating or affective bond of a "we." One could read her politics of natality and its insistence on the move from the natural to the mortal/immortal, from *zoe* to *bios,* as adding another properly political chapter to Hobbes's understanding of the human being. Contrary to Hobbes's position that the rights of human beings are natural, Arendt's political understanding of the human being insists on the transformation of the time of self-interest to the temporality of public happiness and its promise of shared pleasures. This, in turn, allows her to reformulate the solidarity of humanity and its predicament of common responsibility. Whereas Kristeva stresses the mourning that accompanies all gratitude, Arendt sees in the public realm a new kind of pleasure principle.

In her essay *On Violence,* Arendt takes up the issue of whether enlightened self-interest can adequately resolve conflict and prevent violence. Using the example of a rent dispute between tenant and landlord, Arendt argues that "enlightened interest would focus on a building fit for human habitation; however, the argument that '*in the long run* the interest of the building is the *true* interest of both the landlord and the tenant' leaves out of account the time factor which is of paramount importance for all concerned." Because of mortality, she argues, the self qua self cannot calculate in long-term interest: "Self-interest, when asked to yield to 'true interest'— that is, the interest of the world as distinguished from that of the self—will always reply, 'Near is my shirt, but nearer is my skin.' . . . It is the not very noble but adequate response to the time discrepancy between men's private lives and the altogether different life expectancy of the public world." To move from self-interest to "world-interest" requires a move from fear to love of the "public thing" (*OV,* 78)

Love of the public thing occurs only through the vigilant partiality of political friendship that rejects from the outset any notion of truth, engaging instead in the practice of questioning and doubting that marks the secular ordeal of modern humanity: "If the solidarity of mankind is to be based on something more solid than the justified fear of man's demonic capabilities, if the new universal neighborship of all countries is to result in something more promising than a tremendous increase in mutual hatred and a somewhat universal irritability of everybody against everybody else, then a process of mutual understanding and progressing self-clarification on a gigantic scale must take place" (*MDT,* 84). For Arendt, Gotthold Lessing is the figure who embraces this secular ordeal: "He was glad that—to use his parable—the genuine ring, if it had ever existed, had been lost; he was glad for the sake of the infinite number of opinions that arise when men discuss the affairs of this world. If the genuine ring did exist, that would mean an end to discourse and thus to friendship and thus to humanness" (26).[64] Lessing rejoices in that very thing that has caused so much distress—namely, "that the truth once uttered becomes one opinion among many, is contested, reformulated, reduced to one subject of discourse among others" (27). Arendt suggests that Lessing was a "completely political person" because of this understanding of the relation between truth and humanity: "He insisted that truth can exist only where it is humanized by discourse, only where each man says not what just happens to occur to him at the moment, but what he 'deems truth.' But such speech is virtually impossible in solitude; it belongs to an arena in which there are many voices and where the announcement of what each 'deems truth' both links and separates men, establishing in fact those distances between men which together comprise the world" (30–31). This does not amount to toleration; instead, "it has a great deal to do with the gift of friendship, with openness to the world, and finally with a genuine love of mankind" (31).

Lessing's antimony between truth and humanity provides Arendt with a kind of thought experiment. She asks the reader to assume for a moment that the racial theories of the Third Reich could have been proved: "Suppose that a race could indeed be shown, by indubitable scientific evidence, to be inferior; would that fact justify its extermination?" She asks the reader not to make the experiment too easy by invoking a religious or moral principle such as "thou shalt not kill." She does so in order to show a kind of thinking governed by neither legal, moral, nor religious principles, recognizing that none of these principles prevented the worst from happening. This way of thinking without recourse to transcendent principles paradoxically gives rise to a fundamental political principle by which to judge our "truths": "*Would any such doctrine, however convincingly proved, be worth the sacrifice of so much as a single friendship between two men?*" (29).

The political principle is friendship: any doctrine that in principle bars the possibility of friendship must be rejected.[65] Political friendship retreats from a notion of truth as objective. Nonetheless, Arendt argues, it has nothing to do with a kind of subjective relativism in which everything is viewed in terms of the self and its interests. Instead, "it is always framed in terms of the relationship of men to their world, in terms of their positions and opinions." Lessing's understanding of friendship therefore has nothing to do with the warmth of fraternity that desires above all to avoid disputes and conflicts. The excessive closeness of brotherliness, Arendt claims, obliterates all distinctions. And Lessing understood this: "He wanted to be the friend of many men, but no man's brother" (30). While political friendship does not recognize any ultimate arbiter for its disputes and disagreements, it is nonetheless guided by a fundamental exigency: we must assume responsibility for what is just and what is unjust, answering for our deeds and words. Our only remedy for radical evil, Arendt suggests, is these fragile friendships that in the face of humanity's demonic capabilities provide the animating or affective dimension of the solidarity of humanity.[66] These friendships are animated by the willingness to endure the burden of questioning and doubting inherent in the very event of natality itself, characterized at once by a desolation and abandonment that make the banality of radical evil an ever-present threat, even as it allows for the miracle of new beginnings and rebirth.

Arendt argues that the affection that is properly political must have two essential political characteristics—openness to others and plurality. Pleasure and joy, rather than pain and suffering, fulfill these criteria. Raising the question of whether human beings are so shabby that they can only give assistance when spurred and, as it were, compelled by their own pain when they see others suffer, Arendt claims that it is pleasure and not pain that is the intensified awareness of reality. Such pleasure emerges from a passionate openness to the world and love of it, while joy, springing from pleasure in the other, gives rise to dialogue: "In discussing these affects we can scarcely help raising the question of selflessness, or rather the question of openness to others. . . . It seems evident that sharing joy is absolutely superior in this respect to sharing suffering. Gladness, not sadness, is talkative, and truly human dialogue differs from mere talk or even discussion in that it is entirely permeated by pleasure in the other person and what he says" (15). Again, Arendt looks to Montesquieu, for whom joy is also a political feeling springing from the insight that human power is not primarily limited by God or Nature but by the powers of one's equals. "And the joy that springs from that insight, the 'love of equality' which is virtue, comes from the experience that only because this is so, only because there is equality of power, is man not alone. For to be alone means to be without equals" (*EU*, 336). Arendt contrasts this feeling of joy in the company of others with the feelings of anxiety and despair that emerge whenever an individual is rendered impotent, that is, is rendered incapable of acting with others. This impotence, she argues, comes about in situations of complete loneliness.

She agrees with Montesquieu's claim that the situation of complete loneliness is the situation of living under tyranny: "tyranny, based on the essential impotence of all men who are alone, the hubristic attempt to be like God, invested with power individually, in complete solitude" (339).

Our pleasure in the company of others depends on foreignness, a heterogeneity at the very heart of the public space. Arendt understands this early in her thinking but then seems to lose sight of it. The political bond of the "we" is animated by gratitude for the now-shattered origin of natality that makes us forever estranged and foreign and, at the same time, is a source of both pain and joy. Following Arendt's insight that the ethical and political task today is not to resurrect a notion of authority (the loss of which characterizes the modern world) but instead to make the world a "fit dwelling,"[67] we must take seriously Kristeva's suggestion that we first familiarize ourselves with the locus of pain that is given in an original uprooting. We must "inhabit" this archaic dehabitation and primordial separation.[68] If and only if there is gratitude for what remains ineradicably alien and foreign, pleasure rather than grief can be possible—in the company of others. That is the predicament of our common responsibility, which in turn is the guarantor of our right to have rights.

Conclusion: The Political Institution of the Right to Have Rights

> Politics seldom offers ideal or eternal solutions.
> Hannah Arendt, *The Jew as Pariah*

> Philosophy may conceive of the earth as the homeland
> of mankind and one unwritten law, eternal and valid
> for all. Politics deals with men, nationals of many
> countries and heirs to many pasts; its laws are the
> positively established fences which hedge in,
> protect, and limit the space in which freedom
> is not a concept, but a living, political reality.
> Hannah Arendt, *Men in Dark Times*

I conclude with a brief consideration of how Arendt understands the political institution of the right to have rights. Admittedly Arendt's writings on political institutions are limited, although they are not entirely nonexistent.[1] Her essays on post–World War II Europe as well as her extensive writings on Palestine during the 1940s and early 1950s provide many insights. These essays indicate that Arendt's early analyses of the actual political situations of Europe and Palestine are at the very center of her theoretical work. Here I will take up four issues that preoccupy Arendt from her earliest writings that illuminate how she imagines the political institution of the right to have rights: sovereignty, the nation-state, the international, and citizenship rights.

Arendt's essay "Approaches to the 'German Problem,'" written in winter 1945 for the *Partisan Review,* is her first attempt to address the problem of national sovereignty.[2] In chapter 2, we saw that for Arendt the link between national sovereignty and human rights resulted in the decline and eventual disappearance of human rights, especially for those not considered to be nationals. In this early essay, Arendt first begins to articulate this link as she argues against the popular notion of a "German Problem," insisting instead on a "European Problem," the roots of which she locates in the principle of national sovereignty: "It is true, and almost self-evident, that the whole Continent is likely to collapse because of the principle of national sovereignty" (*EU,* 157). Arendt rejects any and all attempts to restore sovereignty to European nation-states, arguing that such efforts are doomed to fail because they are modeled on three premises, all of which are false. First, the restoration effort argues that national sovereignty can be restored if the various nation-states agree on collective security. She argues that such agreement is not possible because of ideological factors. Second, restoration efforts attempt to outline clearly demarcated spheres of interest in order to avert potential clashes between ideological forces. For Arendt, this is a model taken from colonial imperialism that resulted in two world wars and the collapse of the nation-state. Certainly for her this is not a model that ought to be resurrected. Finally, restoration efforts look to "bilateral alliances" to secure national sovereignty. Such alliances are nothing but "nineteen century power politics" which come down to the stronger partner in the alliance dominating the weaker, politically and ideologically. She rejects restoration efforts in any of its three forms, claiming that "restoration promises nothing!" (120).

A second essay written in 1945, "Seeds of an International Fascist Movement" underscores and adds to the concerns of the above-mentioned essay, arguing that "national sovereignty is no longer a working concept of politics, for there is no longer a political organization which can represent or defend a sovereign people, within national boundaries.[3] Thus the 'national state,' having lost its very foundations, leads the life of a walking corpse, whose spurious existence is artificially prolonged by repeated injections of imperialistic expansion" (143). Here Arendt is especially concerned with the rights of refugees, and she is convinced that national sovereignty is unable to address the issue: "Restoration of the European national system means for them rightlessness compared to which the proletarians of the nineteenth century had a privileged status. They might have become the true vanguard of a European movement—and many of them, indeed, were prominent in the Resistance; but they can easily fall prey, also, to other ideologies if appealed to in international terms" (149). Refugees, minority populations, and displaced persons are for Arendt more oppressed than Marx's proletariat; their oppression, taking the form of rightlessness, will continue to prevail unless there is a new understanding of the nation-state and citizenship. With chilling prescience, Arendt predicts that if these populations do not have the opportunity to be part of a reconfigured nation-state, there is the very real possibility of their being enlisted by new international ideological movements. Given the tenacity of national sovereignty with its accompanying nationalism, as well as the imperialism that hides in its shadow, Arendt would not have been surprised by the rise of international terrorist movements. We can speculate that she would agree with thinkers such as Richard Barnet that "international terrorism serves as the successor myth to international communism."[4]

Arendt's essay on a fascist international is extremely important in this context because it cautions against an easy embrace of the "international" as a simple alternative to nationalism. Thus, while James Bohman is correct to point out that Arendt bases her political theory on a notion of the international,[5] we should recognize that for Arendt the notion of the international is not unproblematic. In her analysis of the fascist international, she begins with a discussion of the *Protocols of the Elders of Zion,* pointing out that Franco had the *Protocols* translated into Spanish even though there was no "Jewish problem" in Spain. Why, she asks, do the *Protocols* (which are completely false) provide a model for a fascist international? She answers that it is because they touch on every troubling political theme of our time: its anti-national tenor, its semi-anarchist antagonism to the state, and the idea that the nation-state itself is an outmoded form of political power concentration. She points out that "the motif of global conspiracy in the *Protocols* also corresponded, and still corresponds, to the altered power situation in which, for past decades, politics has been conducted. There are no longer any powers but world powers, and no power politics but global politics. These have been the conditions of modern political life for the past century—conditions, however, to which Western civilization has so far found no adequate response" (143). She goes on to claim that at the core of the fascist international is an anti-national movement that she finds extraordinarily dangerous and that must be avoided in any alternative understandings of the international: "Only when fascism is understood as an anti-national international movement does it become intelligible why the Nazis, with unparalleled coolness, not distracted by national sentimentality or humane scruples as to the welfare of their people, allowed their land to be transformed into a shambles. The German nation has gone down in ruins together with its terrorist regime of twelve years' duration, whose policing apparatus functioned unfailingly until the last minute" (144).

The demise of the German nation in the face of the fascist international leads Arendt to make a distinction between a "global politics" and an "international politics," rejecting the first and embracing the second. Globalization for her is a movement that depends on the collapse of the nation-state. She identifies globalization with an imperialistic politics, which is for her a power politics unlimited in scope and desire. Its anti-national view is rooted in its lack of respect for territorial limit or boundary. In this way, the fascist international is actually a movement that aspires to globalization and dominant world power. As we saw in chapter 2, one of the prerequisites of imperialistic domination for Arendt is the destruction of societies, communities, and nations "whose atomization is one of the prerequisites of [its] domination" and that relies on "modern brands of nationalism [that] are racist to some degree" (207–208). By contrast, the international is not anti-national. For Arendt, the nation lies at the heart of the international, albeit a nation that is no longer nationalistic. Her 1963 essay "Karl Jaspers: Citizen of the World" is helpful insofar as here she argues with Jaspers against globalization in favor of a "world-wide federated political structure" (*MDT,* 84). This structure would be constituted by nations insofar as Arendt defines the political as based on "plurality, diversity, and mutual limitations. A citizen is by definition a citizen among citizens of a country among countries. His rights and duties must be defined and limited, not only by those of his fellow citizens, but also by the boundaries of a territory" (81). While Arendt looks to worldwide international organizations, "without which there can be no lasting peace," she argues that human rights can be achieved only through "regional organization" (*EU,* 156). In the preface to the first edition of *The Origins of Totalitarianism,* Arendt calls for a new political principle "whose validity this time must comprehend the whole of humanity while its power must be strictly limited, rooted in and controlled by newly defined territorial entities" (*OT,* ix). In her earliest essays, Arendt does not shrink from the task of articulating both a new principle of humanity that animates the right to have rights *and* of imagining how the right to have rights would be instituted in new territorial entities and regional organizations.

This is possible for Arendt only if the international community is comprised of nations no longer at risk of developing nationalism: the fundamental task of political theory today is "to find a political principle which would prevent nations from developing nationalism and would thereby lay the fundamentals of an international community capable of presenting and protecting the civilization of the modern world" (*EU,* 207). For Arendt, all forms of nationalism are racist to some degree. We also saw that for Arendt nationalism, with its emphasis on the power of the people, views the state as an enterprise of power: "Aggressive and inclining to expansion, the nation through its identification with the state acquires all these qualities and claims expansion now as a national right, as a necessity for the sake of the nation" (208). Thus, the confrontation with nationalism necessitates an analysis of the origin and legality of state power. For Arendt, the legitimate origin of state power lies in the twin principles of publicity and givenness, themselves inherent in the archaic event of natality. These twin principles demand that power always be based upon right, most fundamentally the right to have rights—that is, the right of a plurality of actors to appear in a public space. Insofar as for Arendt the law is by definition what limits and bounds power, deriving its legal legitimacy also from the fundamental right to have rights, state power never has the legal or political right to aggressive expansion. The state can never claim sovereign power precisely because its power is limited by law.

In her essay *On Violence,* Arendt looks with both approval and indignation to the United States, a state founded upon the principle of law rather than sovereignty but whose practices in the 1960s and 1970s, particularly in Vietnam, violated this founding principle. Arguing in this essay that the violence of the Hobbesian sword will be replaced by the diplomacy of the word only when freedom is no longer linked to state sovereignty, "namely the claim to unchecked and unlimited power," Arendt points to the United States as one of "the few countries where a proper separation of freedom and sovereignty is at least theoretically possible insofar as the very foundations of the American republic would not be threatened by it. Foreign treaties, according to the Constitution, are part and parcel of the law of the land, and—as Justice James Wilson remarked in 1793—'to the Constitution of the United States the term sovereignty is totally unknown'" (*OV,* 107–108). Arendt laments that in the 1960s this founding principle was forgotten and "the American government, for better and for worse, . . . entered into the heritage of Europe as though it were its patrimony—unaware, alas, of the fact that Europe's declining power was preceded and accompanied by political bankruptcy, the bankruptcy of the nation-state and its conception of sovereignty" (108). Arendt is consistent in her condemnation of the linkage of human rights to national sovereignty. Only something like limited sovereignty can ensure human rights rooted in the fundamental right to have rights, which necessarily limits state sovereignty.

In her 1945 essay "On the German Problem," Arendt looks to the European resistance movement for a possible alternative to the restoration of nationalism with its principle of sovereignty: "This restoration, proceeding with the aid of intensified nationalist chauvinist propaganda, particularly in France, is in sharp opposition to the tendencies and aspirations begotten by the resistance movements, which were genuinely European movements" (*EU,* 150). She argues that in its resistance to fascism, the underground movement discovered a positive political principle "which plainly indicated the non-national though very popular character of the new struggle. That slogan was simply EUROPE" (112).[6] Arendt approvingly cites a question raised by the Dutch resistance: "How can we, while preserving cultural autonomy, achieve the formation of larger units in the political and economic field? . . . A good peace is now inconceivable unless the States surrender parts of their economic and political sovereignty to a higher European authority: we leave open the question whether a European Council, or Federation, a United States of Europe or whatever type of unit will be formed" (113). She takes up the slogan of the Dutch underground: *Liberer et fed-*

erer—liberty and federation. The problem of human rights, themselves rooted in the right to have rights, can only be resolved by *not* restoring sovereign rights: "Thus the Dutch contend that 'the problem of equality of rights should not be a matter of restoring sovereign rights to the defeated state but of granting it a limited influence within the European Council or Federation.'" Arendt goes on to cite the French, who have warned that "'essential restrictions on German sovereignty can be envisaged without difficulty only if all the states likewise accept significant limitations on their own sovereignty'" (116). And in a review of Feliks Gross's book *Crossroads of Two Continents: A Democratic Federation of East-Central Europe* that appeared in *Commentary* in 1945–1946, Arendt is in total agreement with the book's "much needed selection from contemporary accounts to show that all the peoples who joined the Resistance did not do so just to fight the German invader but had gotten it into their heads that they were fighting for something. What they were fighting for was a federated Europe" (156). For Arendt, therefore, sovereignty should be replaced with regional federations under the jurisdiction of international institutions that would have only a "pooled" sovereignty that would be limited by international law.

International institutions, most urgently an international criminal court, were for Arendt the only hope humanity has of confronting crimes against humanity. In her postscript to *Eichmann in Jerusalem,* she speculates on what would have happened if Israel had "waived its right to carry out [Eichmann's] sentence once it had been handed down, in view of the unprecedented nature of the court's findings." She speculates that if this had occurred, Israel "might then have had recourse to the United Nations and demonstrated with all the evidence at hand, that the need for an international criminal court was imperative, in view of these new crimes committed against mankind as a whole." She suggests that if Israel had brought Eichmann to the world stage, insisting that his case deserved an international solution, it would have showed the world the need for a permanent international criminal court. Arendt gives a chilling prophecy of the future: "Only by creating, in this way, an 'embarrassing situation' of concern to the representatives of all nations would it be possible to prevent 'mankind from setting its mind at rest' and 'massacre of the Jews . . . from becoming a model for crimes to come, perhaps the small scale and quite paltry example of future genocide.' The very monstrousness of the events is 'minimized' before a tribunal that represents one nation only" (*EJ,* 270).

If the European Resistance movement with its insistence on a federated Europe provides Arendt with the outlines of a positive political alternative to the sovereign nation-state, Zionism provides her with a model of all that is wrong with the continued embrace of nationalism and state sovereignty. Arendt's criticism of and eventual break with the Zionist movement, as well as her reflections on the lost opportunity of what once went by the name of "Palestine," provides further illumination of how she is rethinking, through the right to have rights, the problems of sovereignty, the nation, and citizenship. In 1951, addressing the events in Palestine, Arendt writes: "The birth of a nation in the midst of our century may be a great event; it certainly is a dangerous event. National sovereignty which so long had been the very symbol of free national development has become the greatest danger to national survival for small nations. In view of the international situation and the geographical location of Palestine, it is not likely that the Jewish and Arab peoples will be exempt from this rule" (*JP*, 222). From the early 1940s, Arendt was deeply disturbed by Zionism's nationalism disguised as a democratic state; she harbored no illusions of the complicit relationship between this brand of democracy and ethnic cleansing that has its origin in a concept of nationalism and national sovereignty. Indeed, she saw the Zionist terror in the late 1940s as designed expulsion of the Arab population. In this context, "democracy" comes to mean some vague notion of self-determination, not inalienable human rights for every member of the state. Arendt had tremendous misgivings about modern democracy, not because she was anti-democratic per se but because she saw that democracy in its modern formulation has not broken with sovereignty and at the same time aligns itself with nationalism. This spells disaster for human rights.

Zionism is no exception. While Zionism, particularly from the 1930s onward, viewed itself as working toward a democratic state of Israel, Arendt argues that with its "inherent principle of nationalism and its insistence on absolute sovereignty, [it] can lead only the Balkanization of the whole region and its transformation into a battlefield for the conflicting interests of the great powers to the detriment of all authentic national interests." Arendt advocates a nonnationalist policy in Palestine, arguing for the same kind of federated structure that she supports for Europe: "In the long run, the only alternative to Balkanization is a regional federation which Magnes . . . proposed as long ago as 1943" (217). For her, a "Confederation of Palestine" rather than two national states offered the only hope for preventing continual conflict between the two populations. She proposed a league of nonaggression, mutual defense, and economic cooperation between Arab and Jewish Palestinians.

Finally, in her penultimate essay on the subject, written in 1948, she offers five "objective factors" for a resolution of the Arab-Jewish conflict, which, she argues, "should be axiomatic criteria for the good and the bad, the right and the wrong." The most important factors for Arendt are the first and the last. The first argues for a "Jewish homeland" and not a "pseudo-sovereign Jewish State." The last argues for full political membership for both Jews and Arabs: "Local self-government and mixed Jewish-Arab municipal and rural councils, on a small scale and as numerous as possible, are the only realistic political measures that can eventually lead to the political emancipation of Palestine" (192).

Within the Palestinian Confederation, Arendt sees the opportunity for cultural autonomy and flourishing. For Arendt, the "the real goal of the Jews in Palestine is the building up of a Jewish homeland. This goal must never be sacrificed to the pseudo-sovereignty of a Jewish state" (192). For her, Zionism is primarily about Hebrew culture. Indeed, she suggests that if there is a hope for an alternative to the nationalism, it will lie in understanding "nationalism" not as a political term aligned with the state but instead as "cultural achievement," and not in any narrow sense. She argues that "if nationalism were nothing worse than a people's pride in outstanding or unique achievement, Jewish nationalism would have been nourished by two institutions in the Jewish National Home: the Hebrew University and the collective settlements. Both are rooted in permanent non-nationalist trends in Jewish tradition—the universality and predominance of learning and the passion for justice" (212). For Arendt, these two institutions, the kibbutzim on the one hand, the Hebrew University on the other, supported and inspired "the non-nationalist, anti-chauvinist trend and opposition in Zionism," a trend that Arendt sees as essential to the continued existence of Israel wherein the term "Hebrew citizen" will designate both Arabs and Jews (*JP,* 212–213).

Arendt viewed with dismay and trepidation the creation of the state of Israel in large part because it repeated the nationalist criteria of citizenship that had held sway in the European nation-state, a repetition that promised nothing more than the violence from which Europe was only emerging at the time of Israel's creation. For Arendt, nowhere was the repetition of nationalism more clearly seen than in conditions of Israeli citizenship, and nowhere was the principle of the right to have rights more clearly absent. With the creation of the state of Israel, Arab Israelis are entitled to equality under the law, but only Jewish nationals have immediate citizenship with the material and political advantages that come with it. Only the child or grandchild of a Jew or a convert by a recognized rabbinic authority is a Jew, rendering it impossible for Arab Israelis to be full citizens in the new state. There is no civil marriage in the state, and no rabbi will marry a Jew and a non-Jew. Thus, Arendt saw that Arab citizens in Israel cannot become Israelis with the full political rights that come with such status. Her break with Zionism was due in large part to the creation of a refugee population through the forced expulsion of the Arab population in Israel and the minority status of those Arabs who remained. With chilling foresight based on the history she had just lived through, she predicted nothing but violence for the inhabitants of the new state, a violence perpetrated on the minority population in its midst.

In her review of J. T. Delos's *La Nation,* which appeared in *The Review of Politics* in January 1946, in the context of articulating a "nation-state without nationalism," Arendt gives us the skeleton outline of a new notion of citizenship. She argues that such a state would be an open society that recognizes only citizens, not nationalities, and whose legal order "is open to all who happen to live on its territory" (*EU,* 207).[7] This again suggests that she is defending a notion of citizenship completely severed from national, ethnic, or religious origin, insisting that citizenship be granted to all who live in the state.[8] If we think this last together with her remarks at the end of *On Revolution* in which she understands citizenship in terms of activity, we can conclude that Arendt is allowing for a notion of "active citizenship"—that is, citizenship granted to all who live in and want to participate in the political space of a state or federation. Perfectly consistent with her articulation of the right to have rights, this last principle means that all those who live in a state have the right to political membership—that is, to vote, to seek political office, and to have political representation. It is the *experience* of citizenship, of living and working in a political space, that is the condition for citizenship.

Moreover, and this stems from the principle of givenness animating the right to have rights, for Arendt, refugees, minorities, and all those who have no political identity or political representation must be incorporated into the political space. While Arendt herself gives no suggestions on how to accomplish this, we may begin to speculate on how this might be done. Certainly her insistence on the nation as an "open society" suggests that while she is arguing for "regional territories," these territories themselves must have fluid borders such that there is a fundamental human right to admittance. While there would also be a human right to exit, the practice of denaturalization and denationalization would not be allowed under an international law animated by the right to have rights. As we have seen, no one understood better than Arendt that the first steps leading to the death camps were taken when entire groups of peoples were denaturalized or denationalized, losing the protection of a sovereign legal body. Finally, and this too stems from the principle of givenness inherent to the right to have rights, active citizenship also includes the right to work; without the means to sustain material existence, the right to participate in political affairs is meaningless.

Most significant for a writer more closely (and, as I argued in chapters 2 and 3, mistakenly) identified with the strict separation of the economic and the political of *The Human Condition,* in these early essays, Arendt calls not only for a political federation of nation-states but also for "the combining of one economic system without changing national borderlines" (116). She argues for "a change in the economic system, control of wealth, nationalization, and public ownership of basic resources and major industries" (114). In the review of Delos's *La Nation,* Arendt claims that the federation of Eastern Europe is both a political and an economic necessity.

She embraces economic considerations, denouncing a world in which "everything is decided from the point of view of politics." She goes on to state, "To this new neglect of economic factors on the part of those who make politics must be added the new over-emphasis on power" (157).

In several of her essays on Palestine, Arendt continues her insistence that economic alliances must accompany federated territorial structures. In her last essay on Israel, "Peace or Armistice in the Near East?" written in 1950 after her break with Zionism and the establishment of the state of Israel, she devotes a lengthy section to arguing against the "watertight walls" that separate the Arab and Jewish economic sectors.[9] She is wary of the Israeli policy of not hiring Arab laborers because of Israeli concerns about exploiting Arabs, pointing out that Israel underwent the industrial revolution 150 years after Europe did: "The decisive difference was only that the industrial revolution had created and employed its own fourth estate, a native proletariat, whereas in Palestine the same development involved the importation of workers and left the native population a potential proletariat with no prospect of employment as free laborers" (*JP,* 202). Certainly Arendt is not condoning the exploitation of the Arab laborer. She is sympathetic to the battle of organized Jewish labor against cheap Arab labor and the argument that "Jewish capital ought to avoid the temptation of employing Arabs instead of the more expensive and more rights-conscious Jewish workers." The problem, however, is that what is good in theory may not be good in practice: "To be anti-capitalist in Palestine almost always meant to be practically anti-Arab" (203).

Arendt is extremely critical of the outside financial support of world Jewry that assured that "Jewish-Arab cooperation [would] hardly become an economic necessity for the new Israeli state" (202). In an all-too-brief discussion, she looks to the collective settlements (the kibbutzim) as an alternative to both the exploitation and the neglect of Arab workers, arguing that instead they allowed for "a new form of ownership . . . and new approaches to the troublesome conflicts between city and country, between rural and industrial labor" (214). She ends this essay lamenting that the collective settlements, the only group that actively promoted Jewish-Arab friendship, failed: "It was one of the greatest tragedies for the new State of Israel that these labor elements, notably the Hashomer Hatsair, sacrificed their bi-national program to the *fait accompli* of the United Nations' partition decision" (215).[10]

In her essays on Europe and her essays on Palestine, Arendt gives us clues about how she understands the political institution of the right to have rights. The institution of this fundamental right depends upon the collective limited sovereignty of states, regional federations with open borders, and international institutions, both legal and economic. Perhaps most important of all is a new notion of citizenship whose rightful condition now is the experience of living in and belonging to a particular political space, a belonging that is guaranteed by the inalienable right to active participation in both the economic and political spheres. This new notion of citizenship recognizes that the temporality at work in the political, the space of appearance, is that of natality with its twin principles of *initium* and givenness. For Arendt, the political space is primarily marked by the upsurgence of the new, and therefore our notion of citizenship must be understood as the ongoing activity of redefinition and reformulation of who we are. For Arendt, this ongoing activity of citizenship must be constituted not only by political institutions—themselves founded upon the inalienable right to have rights guaranteed by the event of natality itself—it must also be animated by gratitude and pleasure for the pluralities and singularities that together constitute the public space.

NOTES

INTRODUCTION

1. Seyla Benhabib, *The Reluctant Modernism of Hannah Arendt* (Thousand Oaks, Calif.: Sage Publications, 1996), xxxiii.

2. Ibid., 82. Oddly, in her recent book *The Claims of Culture: Equality and Diversity in the Global Era* (Princeton, N.J.: Princeton University Press, 2002), Benhabib rejects the very framework she criticizes Arendt's thought of lacking; namely, a philosophical universalism grounded in claims about a universal human essence or human nature. She argues instead for a universalism that encompasses three dimensions: moral universalism, the claim that all human beings should be considered as moral equals; legal universalism, the claim that there are certain basic rights that should be given to all human beings that are reflected in legal institutions; and justificatory universalism, rooted in rational normativity with its ideals of impartiality and objectivity. Benhabib's questions to Arendt come back to haunt her: "Why are there certain basic rights that should be given to all human beings?" and "Why should all human beings be considered as moral equals?" Arendt's philosophical account of the right to have rights answers both of these questions. In doing so, Arendt offers a philosophical universalism that is not grounded in a universal human essence or human nature; rather, it is grounded in the universal event of natality.

3. Dana Villa, *Politics, Philosophy, Terror* (Princeton, N.J.: Princeton University Press, 1999), 199.

4. Ibid., 200.

5. Margaret Canovan, *Hannah Arendt: A Reinterpretation of Her Political Thought* (London: Cambridge University Press, 1992), 198–199.

6. Claude Lefort, *Democracy and Political Theory*, translated by David Macey (Minneapolis: University of Minnesota Press, 1988), 54.

7. John Rawls, *The Law of Peoples* (Cambridge, Mass.: Harvard University Press, 1999), 15.

8. Michael Ignatieff, *Human Rights as Politics and Idolatry*, edited and with an introduction by Amy Gutman (Princeton, N.J.: Princeton University Press, 2001), 82. To ground human rights in humanity, Ignatieff claims, is to court a threefold risk: "It puts the demands, needs, and rights of the human species above any other and therefore risks legitimizing an entirely instrumental relation to other species; second, . . . it authorizes the same instrumental and exploitative relationship to nature; and, finally, it lacks the metaphysical claims necessary to limit the human use of human life, in such instances as abortion or medical experimentation" (82–83).

9. Ibid., 83.

10. Ibid. In another passage, citing Elie Wiesel's claim that the Universal Declaration of Human Rights has become the "sacred text of a world-wide secular religion," Ignatieff writes, "Human rights has become the major article of faith of a secular culture that fears it believes in nothing else. It has become the lingua franca of global moral thought, as English has become the lingua franca of the global economy. The question I want to ask about this

rhetoric is this: If human rights is a set of beliefs, what does it mean to believe in it? Is it a belief like a faith? Is it a belief like a hope? Is it something else entirely?" (53). Arendt avoids all forms of belief, whether faith or hope, by providing a philosophical account of the foundation of human rights.

1. THE EVENT OF NATALITY

1. Hannah Arendt, "A Reply to Eric Voegelin," in *Essays in Understanding*, edited by Jerome Kohn (New York: Harcourt Brace and Co., 1994), 403.

2. Arendt is careful to make a distinction between moral or legal responsibility, on the one hand, and political responsibility, on the other. The predicament of common responsibility refers to the political responsibility inherent in the ideal of humanity. See her article "Collective Responsibility" in Hannah Arendt, *Responsibility and Judgment*, edited by Jerome Kohn (New York: Schocken Books, 2003), 150–151. I am indebted to Wesley Swedlow for his insights on the role terror plays in Arendt's notion of humanity, especially those developed in our collaborative keynote presentation at the conference "Collaborations: On Responsibility," Southern Illinois University-Carbondale, April 5, 2002.

3. Ignatieff, *Human Rights as Politics and Idolatry*, 22.

4. Ibid., 20.

5. Ibid., 16.

6. John Rawls, *The Law of Peoples* (Cambridge, Mass.: Harvard University Press, 1999), 15.

7. Ibid., 79.

8. Ibid., 79–80.

9. Ibid.

10. Jürgen Habermas has astutely observed that all Rawls has shown is that "a normative theory of justice of the sort he proposes can gain entry to a culture in which basic liberal convictions are already rooted through tradition and political socializations in everyday practices and in the intuitions of individual citizens." See *Between Facts and Norms: Contributions to a Discourse Theory of Law and Democracy*, translated by William Rehg (Cambridge, Mass.: MIT Press, 1996), 61.

11. Ibid., 24.

12. Ibid., 2–3.

13. Lefort, *Democracy and Political Theory*, 40.

14. Ibid., 40.

15. Ibid., 18.

16. Ibid.

17. Ibid., 19.

18. Ibid., 40.

19. Montesquieu, *The Spirit of the Laws*, translated by Thomas Nugent (1748; reprint, London: Haftner Library of Classics, 1949), Book III.1, 10.

20. See Arendt's essay, "The Eggs Speak Up," in *Essays in Understanding*, specifically 276.

21. See, for example, *On Revolution* (New York: Penguin Books, 1963), 61, and *The Human Condition* (Chicago, Ill: University of Chicago Press, 1958), especially her well-known analysis of the rise of the social, which Arendt argues occurs when life processes—that is, natural biological processes—improperly make their way into the political, thereby signaling the death of the political (see 68–73).

22. Georgio Agamben, *Homo Sacer: Sovereign Power and Bare Life*, trans. Daniel Heller-Roazen (Stanford, Calif.: Stanford University Press, 1998), 126–135. Bernard Flynn makes a similar objection in *Political Philosophy at the Closure of Metaphysics* (Atlantic Highlands, N.J.: Humanities Press, 1992), 193–194.

23. Homi Bhaba makes a similar point in his reading of Benjamin in *The Location of Culture*, which makes all the more surprising his critique of Arendt as offering nothing more than a progressive and repetitive notion of history and action. Arendt's analysis of Benjamin not only supports but offers additional insights to many of Bhaba's claims, especially how the foreign element is introduced into action. Indeed, Arendt's analysis underscores the destructive element at work in citability. To be sure, Bhaba is more interested in how the new is introduced into historical narratives, while I am arguing that in her reading of Benjamin, Arendt is illuminating the temporality at work in natality itself. See Homi Bhaba, *The Location of Culture* (London: Routledge, 1994), 227–229.

24. For an excellent discussion of this point, see Rudolph Gashe, *The Tain of the Mirror* (Cambridge, Mass.: Harvard University Press, 1986), 215.

25. Arendt, *On Revolution,* 106–107.

26. Hans Jonas, *Philosophical Essays: From Ancient Creed to Technological Man* (Chicago: University of Chicago Press, 1974), 139.

27. Ignatieff, *Human Rights as Politics and Idolatry,* 82.

28. In a late essay, *Crisis of the Republic* (New York: Harcourt Brace, 1972), Arendt argues that we must speak of public spaces in the plural.

29. This underscores the importance of promise-making in Arendt's thought, illuminated through the notion of the public space as a space of performative speech. Promise-making, that which binds us to an unpredictable future, is performed. Moreover, it does not require any notion of intentionality on the part of the promise-maker. The intention not to perform the promise does not deprive the speech act of its status as a promise. The promise is still performed, and in its performance it becomes part of the web of relationships that makes up the *inter-esse.*

30. The first reference, found in *Life of the Mind: Thinking* (New York: Harcourt Brace Jovanovich, 1971), is to Heidegger's notion of the *Augenblick* (212). The second reference, found in her essay on Walter Benjamin, is to Heidegger's statement, "Man can speak only insofar as he is the sayer"; Walter Benjamin, *Illuminations,* edited with an introduction by Hannah Arendt (New York: Schocken Books, 1969), 204.

31. Martin Heidegger, *Sein und Zeit* (1927; reprint, Tubingen: Max Niemeyer Verlag, 1979), 374; Heidegger, *Being and Time,* translated by John Macquarrie and Edward Robinson (New York: Harper and Row, 1962), 426.

32. Heidegger, *Being and Time,* 427.

33. Ibid., 238–239.

34. Hannah Arendt, *Men in Dark Times* (New York: Harcourt Brace & Co., 1955), 204.

35. I am indebted to my colleague David F. Krell, who first brought this to my attention in a panel discussion with Niklaus Lagier and myself on the notion of natality at the May 1998 meeting of the Heidegger Conference at Pennsylvania State University.

36. Heidegger, *Being and Time,* 142.

37. Ibid., 110/145.

38. Ibid., 208.

39. See Claude Lefort, "The Death of Immortality?" in *Democracy and Political Theory,* 221–222.

40. See Arendt, *Life of the Mind: Thinking,* 212.

41. In her many readings of this parable, Arendt does show how the gap between the past and the future is made possible because of the insertion of the human being in time, an insertion that breaks up the continuum of time and deflects the present from both the past and the future. This deflected present is what allows for the new. Arendt, however, does not link the event of natality with her account of the linguistic natality.

42. Heidegger, *Being and Time,* 437.

43. Ibid., 437–438. Translation altered.

2. THE PRINCIPLE OF *INITIUM*

1. Thomas Hobbes, *Leviathan,* edited by Richard Tuck (Cambridge: Cambridge University Press, 1991), 91.

2. Ibid.

3. Ibid., 92.

4. Ibid., 45.

5. Ibid.

6. Ibid., 110.

7. Habermas is in complete agreement with Arendt on this point. Hobbes's political theory, he argues, offers the perspective of the participants in the social contract: "a purposive-rational calculation of their own interests." Thus, there is no need to elaborate regulations for a legitimate exercise of political authority. Instead, the sovereign ensures "an ordered system of egoisms that is favored by all the participants anyhow: what appears as morally right and legitimate then issues spontaneously from the self-interested decisions of rational egoists or, as Kant will put it, a 'race of devils.'" See Habermas, *Between Facts and Norms,* 90–91.

8. Rousseau, *Discourse on the Origin of Inequality* (Cambridge: Cambridge University Press, 1997), Preface, Par. 9, 127.

9. This is the basis of Rousseau's disagreement with Locke about the proper ends of government. For Locke, the right to self-preservation carries the right to appropriate in order to preserve. An individual has the right not only to self-preservation but also to the *means* of self-preservation, which is cultivated through labor—namely, property. Therefore, the end of government for Locke is to protect private property. Rousseau disagrees. Because freedom is a higher good than life, the end of government is to protect the freedom of the individual (which will also include his or her self-preservation).

10. Leo Strauss, *Natural Right and History* (Chicago: University of Chicago Press, 1953), 278. Charles Taylor makes a similar point, arguing that Rousseau is the first philosopher of freedom. See Charles Taylor, *Sources of the Self: The Making of Modern Identity* (Cambridge, Mass.: Harvard University Press, 1989), 364.

11. Rousseau, *Discourse on the Origin of Inequality,* Part II, Par. 37, 176.

12. Rousseau, *The Social Contract,* edited by Victor Gourevitch (Cambridge: Cambridge University Press, 2003), Chapter 1, Par. 2, 176.

13. Rousseau, *Discourse on the Origin of Inequality,* in Jean-Jacques Rousseau, *The Discourses and Other Early Political Writings, edited by* Victor Gourevitch (Cambridge: Cambridge University Press, 1997), Part I, Par. 36, 152.

14. Ibid.

15. Rousseau, *Emile,* translated by Allen Bloom (New York: Basic Books, 1979), 235.

16. Ibid., 289.

17. Ibid., 290.

18. Augustine, *The Confessions,* translated by John Ryan (New York: Image Books, 1960), 196.

19. Arendt, *On Revolution.*

20. Herman Melville, *Billy Budd, Sailor* (New York: Penguin, 1986), 297.

21. Lefort, *Democracy and Political Theory,* 39.

22. Here the "fact of birth" is not the event of natality, but the *birth certificate* that counts—the certification of a nationality that imbues one with human rights.

23. Michel Foucault, *Knowledge/Power: Selected Interviews and Other Writings,* edited by Colin Gordon (New York: Pantheon Books, 1980), 121. In this 1976 essay on power, Foucault argues: "What we need, however, is a political philosophy that isn't erected around the

problem of sovereignty, nor therefore around the problems of law and prohibition. We need to cut off the King's head: in political theory that has still to be done." This is a remarkable statement insofar as it suggests that although Foucault certainly has read Arendt, he has not read her closely enough. It is precisely the continuous task of Arendt's political philosophy to cut off the head of the king; that is, to think a notion of power that is not linked to a notion of sovereignty and furthermore to think a notion of the law that is not understood as prohibition.

24. Luc Ferry, *Rights—The New Quarrel between the Ancients and the Moderns*, translated by Franklin Philip (Chicago: University of Chicago, 1990), 21.

25. Ignatieff, *Human Rights as Politics and Idolatry*, 5.

26. Ibid., 57.

27. Ibid.

28. Ibid., 89.

29. Arendt, "The Eggs Speak Up," 279.

30. Habermas, *Between Facts and Norms*, 82–83.

31. Ibid., 82.

32. Ibid., 33.

33. Ibid., 32.

34. Ibid., 3.

35. Ibid., 104.

36. Ibid., 448.

37. Ibid.

38. Ibid., 449.

39. Ibid., 512.

40. Ibid., 456.

41. Ibid., 512.

42. Hannah Arendt, *On Violence* (New York: Harcourt, Brace & World, 1970), 44.

43. Amartya Sen, *Resources, Values, and Development* (Cambridge: Harvard University Press, 1984), 317.

44. Ibid., 324.

45. In his response to Ignatieff's essay, Thomas Laqueur seems to continue this opposition between right and power, arguing that the "fratricidal war in Sri Lanka is not varying views of human rights but a chasm between their respective views of the state; and what is needed to effect a cessation of atrocities is probably the general acceptance not of a theory of rights but of a common view of power and its exercise." Quoted in Ignatieff, *Human Rights as Politics and Ideology*, 137. Arendt's claim is that political rights and the exercise of political power are inseparable.

46. Sen, *Resources, Values, and Development*, 314–317.

47. Ibid., 323.

48. Henry Shue, *Basic Rights: Subsistence, Affluence, and U.S. Foreign Policy* (Princeton, N.J.: Princeton University Press, 1996), 18.

49. Bhaba, *The Location of Culture*, 190–192.

50. Ibid., 190.

51. Arendt, *Essays in Understanding*, 428–447.

52. Arendt, *Men in Dark Times*, ix.

3. THE PRINCIPLE OF GIVENNESS

1. Hannah Arendt, "Philosophy and Politics," *Social Research* 57, no. 1 (Spring 1990), 103.

2. This letter, written in July 1963, is Arendt's response to Scholem's letter of June 1962 in which he roundly criticizes her book *Eichmann in Jerusalem*. In his letter, he writes that the tone of *Eichmann in Jerusalem* displays a lack of *Herezenstakt*, a warm feeling toward her fellow Jews. See Hannah Arendt, *The Jew as Pariah*, 242.

3. See Hannah Arendt, *Eichmann in Jerusalem: A Report on the Banality of Evil* (New York: Penguin Books, 1963), 269, 273.

4. Augustine, *On the Trinity*, translated by Stephen McKenna (Washington, D.C.: Catholic University Press, 1963), 1.12.26.

5. Ibid., 5.5.6.

6. Ibid., 7.2.3.

7. Ibid., 9.4.4.

8. Ibid., 11.2.2.

9. Augustine, *Free Choice of the Will*, translated by Anna S. Benjamin and L. H. Hackstaff (New York: Macmillan, 1989), III.6.64.

10. Monica is the exemplary figure of Kristeva's claim of the centrality of the maternal psyche in its singularity and uniqueness in the spiritual transformation of the self: "To transform the nascent being into a speaking and thinking being, the maternal psyche takes the form of a passageway between *zoe* and *bios*, between physiology and biography, between nature and spirit." And as Jean Bethke Ehlstein points out, Augustine repeatedly uses "the metaphor of laboring and giving birth to describe his own activities in trying to bring forth a transformed way of understanding his own self." *Augustine and the Limits of Politics* (Notre Dame, Ind.: Notre Dame University Press, 1995), 47. The *Confessions* repeatedly suggest that is only through the maternal that these birth pangs and transformation are possible. For example, in Book V, Augustine writes: "Words cannot describe how dearly she loved me or how much greater was the anxiety she suffered for my spiritual birth than the physical pain she endured in bringing me into the world." And in Book IX, reflecting on the death of his mother, he writes: "In the flesh she brought me to birth in this world: in her heart she brought me to birth in your eternal light."

11. Augustine, *The Confessions*, IX.11.2.

12. In *Homo Sacer*, Giorgio Agamben writes, "Arendt establishes no connection between her research in *The Human Condition* and the penetrating analysis she had previously devoted to totalitarian power (in which a biopolitical perspective is altogether lacking)." *Homo Sacer: Sovereign Power and Bare Life*, translated by Daniel Heller-Roazen (Stanford: Stanford University Press, 1998), 4. I disagree: Arendt's analysis of the racialization of the Jewish people is an extended examination of the ways in which biopolitics is at work in the modern political space. I would also argue that her critique of the political emancipation of the bourgeoisie—liberation of the economic in the political—can be read as a biopolitics.

13. Agamben, *The Coming Community*, translated by Michael Hardt (Minneapolis: University of Minnesota Press, 2005), 96.

14. Ibid., 91.

15. Kristeva's analysis of this dimension of the "we" in *Strangers to Ourselves* as well as in *Hannah Arendt* illuminates Arendt's analysis on this point. She writes: "There will be a possible *we* thanks only to that splitting that all wanderers are urged to discover within themselves and in others, after they first recognized themselves in Christ. Paul is not only a politician. He is a psychologist, and if the institution he sets up is also political, its efficiency rests on the psychological intuition of its founder. It is based on the logic of desire in which one is led to identify with a splitting that henceforth is no longer dangerously set . . . but is, thanks to Christ, experienced as a transition toward a spiritual liberation starting from and within the concrete body" (82).

16. Agamben, *Homo Sacer*, 102.

17. Ibid., 65.

18. Ibid., 65–66.

19. Here Kristeva's analysis of *caritas* illuminates Arendt's claim. In *Strangers to Ourselves,* she writes: "Differences in love are not to be erased but forgiven. . . . Differences between the worthy and the unworthy, the faithful and the unfaithful, the good and the bad—and even the heretics; those are not to be reconciled but brought together though the possibility of giving and the acceptance of what is given. The pilgrim gives and receives, his wandering having become gift is an enthusiasm: it is known as *caritas*" (translated by Leon S. Roudiez [New York: Columbia University Press, 1992], 84).

20. Agamben, *Homo Sacer,* 134.

21. Robert Antelme, *The Human Race,* translated by Jeffrey Haight and Annie Mahler (Evanston, Ill.: Northwestern University Press, 1992), 5.

22. Ibid., 219.

23. The *Habilitationschrift* is the second dissertation a student at a German University must write in order to secure a teaching position at the university level.

24. Arendt is indebted to Heidegger's notion of *Geworfenheit*—the facticity into which we are thrown at birth.

25. See Richard Bernstein, *Hannah Arendt and the Jewish Question* (Cambridge, Mass.: MIT Press, 1996), 48. See also his essay "Hannah Arendt's Zionism?" in *Hannah Arendt in Jerusalem,* edited by Steven E. Aschheim (Berkeley: University of California Press, 2001).

26. Given her insistence on the *political* right of everydayness, it is strange that Arendt still continues to be read as a thinker who embraces a heroic notion of the public space. As her reading of Kafka indicates, this is entirely incorrect. If one pays close attention to her understanding of the hero in *The Human Condition,* the book that serves as the basis for so much of this misreading, Arendt is clear that the hero is not one who has performed some glorious deed: "The hero the story discloses needs no heroic qualities; the word 'hero' originally, that is, in Homer, was no more than a name given each free man who participated in the Trojan enterprise and about whom a story could be told" (*Human Condition,* 186). The hero is one who has really lived his or her life such that a real story can be told: "The real story in which we are engaged as long as we live has no visible or invisible maker because it is not made up. The only 'somebody' it reveals is its hero" (ibid.). We could read Arendt's biography of Rahel Varnhagen as a story that in the end is real, because Rahel finally acknowledges who she is and becomes the heroine of her story. Rahel's acknowledgment of her Jewishness permits Arendt to tell a true story and not a fiction. See *Rahel Varnhagen, The Life of a Jewish Woman,* translated by Richard and Clara Winston (New York: Harcourt Brace Jovanovich, 1974).

27. It is the inability of Jews to perform mundane, everyday tasks, such as shopping for milk and bread, that underscores their exclusion from public life: "Once we could buy our food and ride in the subway without being told we were undesirable. . . . We try the best we can to fit into a world where you have to be sort of politically minded when you buy your food"; *The Jew as Pariah,* edited by Victor Gourevitch (Cambridge: Cambridge University Press, 2003), 60–61.

28. Orlando Patterson, *Slavery and Social Death: A Comparative Study* (Cambridge, Mass.: Harvard University Press, 1982), 5.

29. Ibid., 2, emphasis mine.

30. For an especially good analysis of Arendt's critique of the Zionist dream of a nationalist state, see Bat-Ami Bar On, *The Subject of Violence: Arendtian Exercises in Understanding,* especially the sixth chapter "Violence in the Intersection of Nationalism and the State Form" (Lanham, Md.: Rowman & Littlefield Publishers, 2002).

31. This is why, for Arendt, forgiveness is necessary for action. The capacity for initiating something new carries unknown consequences for oneself and for the web into which

one is inserted. Thus, Arendt argues that we need the "constant mutual release" forgiveness provides in order to be willing to begin again. Arendt, *Human Condition*, 249.

32. See Bhaba, *The Location of Culture*, 70.

33. See Benhabib, *The Claims of Culture*, 94.

4. THE PREDICAMENT OF COMMON RESPONSIBILITY

1. Readers of Arendt all too often overlook this aspect of her thought, which leads to an overly optimistic view of Arendt's understanding of political action. For such optimistic readings of Arendt's notion of action, see Patricia Bowen-Moore, *Hannah Arendt's Philosophy of Natality* (New York: St. Martin's Press, 1989); Maurizio Passerin D'Entreves, *The Political Philosophy of Hannah Arendt* (London: Routledge, 1994); and Jacques Taminiaux, *The Thracian Maid and the Professional Thinker* (New York: State University of New York Press, 1997). For a more cautious analysis of Arendt's notion of action, one that insists on taking into account the negative side of action—that is, violence, see John McGowan, "Must Politics Be Violent? Arendt's Utopian Vision," in *Hannah Arendt and the Meaning of Politics*, edited by Craig Calhoun and John McGowan (Minneapolis: University of Minnesota Press, 1997). McGowan's account of violence, however, does not consider Arendt's understanding of evil. Indeed, he argues that "Arendt consistently refused throughout her career to attempt any explanation of evil while persistently calling our attention to the relevance of its existence as a political fact" (269). In his later analysis, *Hannah Arendt: An Introduction* (Minneapolis: University of Minnesota Press, 1998), McGowan does give an account of Arendt's understanding of evil but views it entirely through the Arendtian lens of thinking and judging, neglecting altogether her notion of radical evil. While I do not disagree that Arendt's later analyses of thinking and judging add to our understanding of evil, I want to argue that her understanding of the banality of evil is rooted in her account of radical evil and the radical superfluousness of the human being, superfluousness that itself can only be understood through Arendt's account of natality. Steven Aschheim's work on Arendt (*In Times of Crisis: Essays on European Culture, Germans and Jews* [Madison: University of Wisconsin Press, 2001]) is to my mind the least optimistic about action as the promise of new beginnings, arguing that Arendt's analysis of radical evil rejects an understanding of evil in terms of particular national and historical categories, instead favoring more general historical and psychological categories. He implicitly suggests that Arendt's insight into the psychological roots of evil would yield a far less optimistic reading of Arendt's notion of action. Aschheim, however, does not develop Arendt's psychological insights.

2. For a groundbreaking analysis of how Kristeva's notion of abjection provides crucial light in illuminating Arendt's thought, see Norma Claire Moruzzi, *Speaking through the Mask: Hannah Arendt and the Politics of Social Identity* (Ithaca, N.Y.: Cornell University Press, 2000). Moruzzi's analysis focuses on the ways in which Arendt's political thought attempts to exclude the abject from political life while at the same time recognizes the force of the abject in her analysis of the worldly achievement of artifice and her understanding of performance as requiring the actor to assume the masquerade of individual self-presentation. Moruzzi devotes an entire chapter to an analysis of the banality of evil, arguing that it is rooted in the thoughtless refusal of this masquerade (see 114–115.) She ends her analysis with an emphasis on the hopefulness inherent in new beginnings. My focus in this chapter is to render problematic the promise of beginning granted in the event of natality; I argue that the event of natality carries its abjection with it and that thereby the promise of beginning is tempered by the threat of radical evil.

3. We must be careful, therefore, not to jump to the conclusion that Arendt changes her mind on radical evil. She agrees with Jaspers that radical evil cannot be attributed to a demonic nature. Later in her exchange with Gershom Scholem, she argues that evil is not

radical if by radical one means "deep"; Arendt, *The Jew as Pariah,* 251. Rather, she writes, evil is like a fungus that spreads on the surface of things. That does not make it any less radical or horrible, however. Arendt's use of the metaphor of fungus indicates that she disagrees with Kant's argument that radical evil has a *root* in human nature.

4. See also "Social Science and Concentration Camps" in *Essays in Understanding.* Arendt writes, "The concentration camps are the laboratories in the experiment of total domination, for human nature being what it is, this goal can be achieved only under the extreme circumstances of a *human-made hell*" (240, emphasis mine).

5. See Lefort, *Democracy and Political Theory.*

6. Arendt, *Hannah Arendt/ Karl Jaspers: Correspondence, 1926–1929,* edited by Lotte Kohler and Hans Saner, translated by Robert and Rita Kimber (New York: Harcourt Brace Jovanovich, 1985), Letter 109, 166.

7. Ibid.

8. In *Emile,* Rousseau argues that pity is the move from the *amour de soi* to the *amour propre. Amour de soi,* the level of need, becomes the *amour propre* through the awakening of desire in which the sentiment of pity becomes socialized. To turn pity back on the self is to move from desire back to need. In Eichmann's case, this has the effect of a "post-desire" need, which explains why he is able so easily to give up his desire.

9. Agamben, *Homo Sacer,* 28–29.

10. See Jean-Luc Nancy, "Abandoned Being," in *The Birth to Presence* (Stanford, Calif.: Stanford University Press, 1993), 43–44.

11. Lyotard argues that Arendt subverts her most important psychoanalytic insight that links the terror of totalitarianism to a need that is born out of the fear of natality with its inherent new beginning by a historical-political point of view that refuses to acknowledge its own psychological underpinnings. See Jean-Francois Lyotard, *Toward the Post-Modern* (Atlantic Highlands, N.J.: Humanities Press, 1993), especially his essay on Arendt, "The Survivor," page 56. I suggest that in *The Origins of Totalitarianism,* Arendt does develop the psychoanalytic analysis of the relationship of terror to the fear and rejection of our finitude, which includes the fear and rejection of natality with its unpredictability, although I agree with Lyotard that she does not go nearly far enough. Kristeva's analysis of Arendt insists on developing the very points that Arendt leaves undeveloped. See also Steven A. Aschheim, *In Times of Crisis: Essays on European Culture, Germans, and Jews* (Madison: University of Wisconsin Press, 2001), especially chapter 11, "Nazism, Culture, and *The Origins of Totalitarianism.*" I am indebted to Aschheim's book for first bringing to my attention Lyotard's work on Arendt.

12. Julia Kristeva, *Hannah Arendt,* translated by Ross Guberman (New York: Columbia University Press, 2001), 128.

13. Ibid., 180.

14. Ibid., 180–181.

15. Ibid., 180.

16. Ibid.

17. Julia Kristeva, *Tales of Love,* translated by Leon S. Roudiez (New York: Columbia University Press, 1987), 127.

18. Julia Kristeva, *Powers of Horror: An Essay on Abjection,* translated by Leon S. Roudiez (New York: Columbia University Press, 1982), 10.

19. Ibid., 9–10.

20. Julia Kristeva, *Melanie Klein,* translated by Ross Guberman (New York: Columbia University Press, 2001), 28–29.

21. Ibid., 158.

22. Ibid., 141.

23. Ibid.

24. Hobbes, *Leviathan,* 31.

25. Kristeva argues that our only hope of preventing future Columbines is the articulation of our fantasies; that is, to put "our desire for death into words." This is possible because our desire for death is already imbued with a primordial symbolism. See *Melanie Klein*, 244–245.

26. Sigmund Freud, *Three Essays on a Theory of Sexuality*, translated by James Strachey (New York: Basic Books, 1975), 2.

27. Kristeva, *Melanie Klein*, 148–149.

28. Ibid., 156.

29. Ibid., 142.

30. Ibid., 161.

31. Ibid., 86. Kristeva's analysis of sadism is the inversion of her analysis of phobia in *Powers of Horror*, where she posits that in phobia, fear and aggressivity come back to the individual from the outside. Instead of entertaining the fantasy of devouring the mother, however, the phobic subject fantasizes that he or she is being devoured.

32. Jessica Benjamin, *Bonds of Love* (New York: Pantheon Books, 1988), 67.

33. Kristeva, *Melanie Klein*, 91.

34. Kristeva points out that envy must be distinguished from jealousy, where one does have the object but fears losing it.

35. Kristeva, *The Kristeva Reader*, edited by Toril Moi (New York: Columbia University Press, 1986), 95.

36. Kristeva, *Powers of Horror*, 197.

37. Kristeva, *Hannah Arendt*, 138.

38. Ibid., 181.

39. Kristeva, *Powers of Horror*, 39.

40. Ibid., 25.

41. Ibid., 35.

42. Ibid., 178. Céline wrote four pamphlets supporting fascist ideology: *Mea Culpa* (1936), *Bagatelles pour un massacre* (1937), *L'Ecole des cadavers* (1938), and *Les Beaux Draps* (1941).

43. Ibid., 179.

44. Ibid., 23.

45. Ibid., 147.

46. Kristeva, *Melanie Klein*, 84–85.

47. Ibid., 76.

48. Ibid., 79.

49. Ibid., 188.

50. While it is beyond the scope of this work, a complete account of gratitude and reparation would show how reparation works toward a translation of grief into an exile into the symbolic.

51. Kristeva, *Strangers to Ourselves*, 40.

52. Ibid., 13.

53. Kristeva, *Melanie Klein*, 24.

54. Ibid., 89.

55. Ibid., 241.

56. Kristeva, *Hannah Arendt*, 181.

57. For a detailed analysis of Arendt's understanding of the hypocrite, see Moruzzi, *Speaking through the Mask*, 32–37. Confining her analysis to Arendt's *On Revolution*, Moruzzi emphasizes the hypocrite's refusal to understand the self as artifice, a multiple and changing appearance among a multiplicity of appearances. While entirely in agreement with Moruzzi's reading of the hypocrite in *On Revolution*, I suggest that in *Life of the Mind*, Arendt adds significantly to her own understanding of the hypocrite by introducing the hypocrite's broken promise to pleasure.

58. Kristeva, *Hannah Arendt*, 47.

59. Ibid., 140.

60. See, for instance, the conclusion to her "Reply to Eric Voegelin" in *Essays in Understanding*, 408.

61. Because she insists on an affective dimension to political life, Arendt is in the tradition of Montesquieu, who argues that the laws and institutions (the form) of any political regime are always animated by an affective principle (the spirit of the laws) that establishes the political bond. For Montesquieu's argument, see *Spirit of the Laws*, especially Part I. For Arendt's reading of Montesquieu on this point, see "On The Nature of Totalitarianism," in *Essays in Understanding*, 331–333.

62. Kristeva, *Hannah Arendt*, 180–181, emphasis mine.

63. In the same essay that Kristeva refers to ("Religion and Politics"), Arendt argues that the remedy to totalitarian evil is to embrace the doubt that characterizes the modern secular world instead of believing in heaven or fearing hell.

64. Arendt refers to Lessing in the context of accepting the Lessing Prize of the Free City of Hamburg in 1951. Her address, "On Humanity in Dark Times," is a self-conscious reflection on the difficulty she faces as a German Jew who returns from forced exile to accept a humanitarian award from the country that forced her to flee.

65. For an extended analysis of Arendt, Lessing, and the ways in which friendship might provide a bulwark against radical evil, see Lisa J. Disch, "On Friendship in 'Dark Times,'" in *Feminist Interpretations of Hannah Arendt*, edited by Bonnie Honig (University Park: Pennsylvania State University, 1995). Disch and I do not disagree on the centrality of friendship for confronting the evil of totalitarianism. Disch's analysis, however, concentrates on how given identities can be challenged by the "vigilant partisanship" of friendship. My focus differs from Disch's analysis in that I argue that political friendship is the affective principle that animates the political bond, or what Arendt calls the "solidarity of humanity." The "predicament of common responsibility," in which this solidarity is both terrifying and unifying, is able to be borne through this type of friendship.

66. Following Montesquieu, I want to emphasize that in arguing for political friendship as a remedy for radical evil, it is also the case that this affective dimension of political life cannot be divorced from the institutions and laws that form governments. It is outside the scope of this work to raise the further question, "What would the form of institutions and laws look like if animated by these fragile friendships that insist on the burden of questioning and doubt?" While this may seem to provide a weak remedy to radical evil, it seems to me that such weakness or lack of guarantees is endemic to what Arendt calls the fragility of human affairs, which is, after all, the fragility of natality.

67. See Hannah Arendt, "What Is Authority?" in *Between Past and Future* (New York: Penguin), 141.

68. Kristeva, *Melanie Klein*, 196.

CONCLUSION

1. Thus, scholars such as Ernst Gellner and Steven E. Aschheim overstate the case against Arendt's lack of concern with the everyday world of politics and the workings of representative liberal democracy. Both these scholars overlook Arendt's essays in the 1940s and 1950s. They also overlook Arendt's engagement with modern democracy through her many reflections on the nation-state. Arendt finds modern democracy to be inseparable from the twin principles of nationality and sovereignty, which she finds extremely problematic precisely because these two principles prevent the modern democratic state from being truly representative of all its citizens. See Ernst Gellner, "From Konigsberg to Manhattan (Or Hannah, Rahel, Martin, and Elfriede or Thy Neighbor's *Gemeinschaft*)," in his

Culture, Identity, and Politics (Cambridge: Cambridge University Press, 1987); and Steven E. Aschheim, "Nazism, Culture, and *The Origins of Totalitarianism*," in his *In Times of Crisis: Essays on European Culture, Germans, and Jews* (Madison: University of Wisconsin Press, 2001).

2. Hannah Arendt, "Approaches to the 'German Problem,'" in *Essays in Understanding, 1930–1954,* edited by Jerome Kohn (New York: Harcourt Brace & Company, 1994); first published in *Partisan Review* XII, no. 1 (Winter 1945).

3. Hannah Arendt, "The Seeds of a Fascist International," in *Essays in Understanding, 1930–1954,* edited by Jerome Kohn (New York: Harcourt Brace & Company, 1994); first published in *Jewish Frontier,* June 1945.

4. See Richard J. Barnet, "The Terrorism Trap," *The Nation,* December 2, 1996, 18–21. See also his *Global Reach: The Power of the Multinational Corporations* (London: Jonathan Cape, 1974).

5. James Bohman, "The Moral Costs of Political Pluralism: The Dilemmas of Difference and Equality in Arendt's 'Reflections on Little Rock,'" in *Hannah Arendt: Twenty Years Later,* edited by Larry May and Jerome Kohn (Cambridge: Massachusetts Institute of Technology Press, 1996).

6. Capitalization in the original text.

7. Arendt's review of Delos's *La Nation* (2 vols. [Montreal: Editions de l'Arbre, 1944]) appeared in *The Review of Politics* VIII, no. 1 (January 1946).

8. Bernard Avishai's article "Saving Israel from Itself: A Secular Future for the Jewish State" (*Harpers* [January 2005]) points to a landmark petition brought before Israel's High Court of Justice in December 2004. The petitioners, Arabs born in Israel, asked the court to order the Ministry of Interior to register them as "Israeli" in the Registry of Population. Avishai commented: "The petitioners are asking the state to recognize an inclusive, earned formed of nationality, coterminous with and redundant to citizenship" (42). This petition underscores Arendt's insistence that nationality be severed from citizenship. For Arendt, the earned form of nationality would be recast as the "earned form of citizenship," earned by the very experience of Israel itself, an experience of *both* Israeli Arabs and Jews.

9. Arendt, *The Jew as Pariah,* 201.

10. We can speculate that Arendt would view with great concern the building of a wall that is creating Palestinian enclaves separated by highways and bypass roads from the rest of Israel, because of which Palestinians face the very real prospect of owning businesses that are not viable.

WORKS CITED

Agamben, Giorgio. *The Coming Community.* Translated by Michael Hardt. Minneapolis: University of Minnesota Press, 2005.

———. *Homo Sacer: Sovereign Power and Bare Life.* Translated by Daniel Heller-Roazen. Stanford: Stanford University Press, 1998.

Antelme, Robert. *The Human Race.* Translated by Jeffrey Haight and Annie Mahler. Evanston, Ill.: Northwestern University Press, 1992.

Arendt, Hannah. *Between Past and Future: Eight Exercises in Political Thought.* 1961; reprint, New York: Viking Press, 1968.

———. *Crises of the Republic.* New York: Harcourt Brace, 1972.

———. *Eichmann in Jerusalem: A Report on the Banality of Evil.* New York: Penguin Books, 1963.

———. *Essays in Understanding.* Edited by Jerome Kohn. New York: Harcourt Brace and Co., 1994.

———. *Hannah Arendt/ Karl Jaspers: Correspondence, 1926–1929.* Edited by Lotte Kohler and Hans Saner. Translated by Robert and Rita Kimber. New York: Harcourt Brace Jovanovich, 1985.

———. *The Human Condition.* Chicago, Ill.: University of Chicago Press, 1958.

———. *The Jew as Pariah: Jewish Identity and Politics in the Modern Age.* Edited by Ron H. Feldman. New York: Grove Press, 1978.

———. *Lectures on Kant's Political Philosophy.* Edited and with an interpretive essay by Ronald Beiner. Chicago: University of Chicago Press, 1982.

———. *Life of the Mind.* Vol. I, *Thinking.* New York: Harcourt Brace Jovanovich, 1978.

———. *Life of the Mind.* Vol. II, *Willing.* New York: Harcourt Brace Jovanovich, 1978.

———. *Love and Saint Augustine.* Edited by Joanne Vecchiarelli Scott and Judith Chelius Stark. Chicago, Ill.: University of Chicago Press, 1996.

———. *Men in Dark Times.* San Diego, New York, London: Harcourt Brace & Co., 1968.

———. "Philosophy and Politics." *Social Research* 57, no. 1 (Spring 1990): 73–103.

———. *Rahel Varnhagen, The Life of a Jewish Woman.* Translated by Richard and Clara Winston. New York: Harcourt Brace Jovanovich, 1974.

———. *The Origins of Totalitarianism.* New York: Harcourt, Brace, 1952.

———. *On Revolution.* New York: Penguin Books, 1963.

———. *On Violence.* New York: Harcourt, Brace & World, 1970.

———. *The Promise of Politics.* Edited by Jerome Kohn. New York: Schocken Books, 2005.

———. *Responsibility and Judgment.* Edited by Jerome Kohn. New York: Schocken Books, 2003.

Aschheim, Steven E. *In Times of Crisis: Essays on European Culture, Germans, and Jews.* Madison: University of Wisconsin Press, 2001.

Augustine. *The Confessions.* Translated by John Ryan. New York: Image Books, 1960.

———. *Free Choice of the Will.* Translated by Anna S. Benjamin and L. H. Hackstaff. New York: Macmillan, 1989.

————. *On the Trinity.* Translated by Stephen McKenna. Washington, D.C.: Catholic University Press, 1963.

Avishai, Bernard. "Saving Israel from Itself: A Secular Future for the Jewish State." *Harpers,* January 2005.

Bar On, Bat-Ami. *The Subject of Violence: Arendtian Exercises in Understanding.* Lanham, Md.: Rowman & Littlefield, 2002.

Barnet, Richard J. *Global Reach: The Power of the Multinational Corporations.* London: Jonathan Cape, 1974.

————. "The Terrorism Trap." *The Nation,* December 2, 1996, 18–21.

————, with John Cavanaugh. *Global Dreams: Imperial Corporations and the New World Order.* New York: Simon and Shuster, 1995.

Benhabib, Seyla. *The Reluctant Modernism of Hannah Arendt.* Thousand Oaks, Calif.: Sage Publications, 1996.

Benjamin, Jessica. *Bonds of Love.* New York: Pantheon Books, 1988.

Benjamin, Walter. *Illuminations.* Edited and with an introduction by Hannah Arendt. New York: Schocken Books, 1969.

Bernstein, Richard. *Hannah Arendt and the Jewish Question.* Cambridge, Mass.: MIT Press, 1996.

————. "Hannah Arendt's Zionism?" In *Hannah Arendt in Jerusalem.* Edited by Steven E. Aschheim. Berkeley: University of California Press, 2001.

Bhaba, Homi. *The Location of Culture.* London: Routledge, 2004.

Bohman, James. "The Moral Costs of Political Pluralism: The Dilemmas of Difference and Equality in Arendt's 'Reflections on Little Rock.'" In *Hannah Arendt: Twenty Years Later.* Edited by Larry May and Jerome Kohn. Cambridge, Mass.: MIT Press, 1996.

Bowen-Moore, Patricia. *Hannah Arendt's Philosophy of Natality.* New York: St. Martin's Press, 1989.

Canovan, Margaret. *Hannah Arendt: A Reinterpretation of Her Political Thought.* London: Cambridge, 1992.

D'Entreves, Maurizio Passerin. *The Political Philosophy of Hannah Arendt.* London: Routledge, 1994.

Disch, Lisa J. "On Friendship in 'Dark Times.'" In *Feminist Interpretations of Hannah Arendt.* Edited by Bonnie Honig. University Park: Pennsylvania State University, 1995.

————. *Hannah Arendt and the Limits of Philosophy.* New York: Cornell University Press, 1994.

Elshtain, Jean Bethke. *Augustine and the Limits of Politics.* Notre Dame, Ind.: Notre Dame University Press, 1995.

Ferry, Luc. *Rights—The New Quarrel between the Ancients and the Moderns.* Translated by Franklin Philip. Chicago, Ill.: University of Chicago, 1990.

Flynn, Bernard. *Political Philosophy at the Closure of Metaphysics.* Atlantic Highlands, N.J.: Humanities Press, 1994.

Foucault, Michel. *Knowledge/Power: Selected Interviews and Other Writings.* Edited by Colin Gordon. New York: Pantheon Books, 1980.

————. "The Subject and Power." In Herbert Dreyfus and Paul Rabinow, *Michel Foucault, Beyond Structuralism and Hermeneutics.* Chicago, Ill.: University of Chicago Press, 1982.

Freud, Sigmund. *Three Essays on a Theory of Sexuality.* Translated by James Strachey. New York: Basic Books, 1975.

Gashe, Rodolph. *The Tain of the Mirror.* Cambridge, Mass.: Harvard University Press, 1986.

Gellner, Ernst. *Culture, Identity, and Politics.* Cambridge: Cambridge University Press, 1987.

Habermas, Jürgen. *Between Facts and Norms: Contributions to a Discourse Theory of Law and Democracy.* Translated by William Rehg. Cambridge, Mass.: MIT Press, 1996.

Heidegger, Martin. *Being and Time.* Translated by John Macquarrie and Edward Robinson. New York: Harper and Row, 1962.

Hobbes, Thomas. *Leviathan.* Edited by Richard Tuck. Cambridge: Cambridge University Press, 1991.

Ignatieff, Michael. *Human Rights as Politics and Idolatry.* Edited and with an introduction by Amy Gutman. Princeton, N.J.: Princeton University Press, 2001.

Jonas, Hans. *Philosophical Essays: From Ancient Creed to Technological Man.* Chicago, Ill.: University of Chicago Press, 1974.

Kateb, George. *Hannah Arendt: Politics, Conscience, Evil.* Oxford: Martin Robertson, 1984.

Kristeva, Julia. *Hannah Arendt.* Translated by Ross Guberman. New York: Columbia University Press, 2001.

———. *The Kristeva Reader.* Edited by Toril Moi. New York: Columbia University Press, 1986.

———. *Melanie Klein.* Translated by Ross Guberman. New York: Columbia University Press, 2001.

———. *Powers of Horror: An Essay on Abjection.* Translated by Leon S. Roudiez. New York: Columbia University Press, 1982.

———. *Strangers to Ourselves.* Translated by Leon S. Roudiez. New York: Columbia University Press, 1992.

———. *Tales of Love.* Translated by Leon S. Roudiez. New York: Columbia University Press, 1987.

Laqueur, Thomas. "The Moral Imagination and Human Rights." In Michael Ignatieff, *Human Rights as Politics and Ideology.* Edited and with an introduction by Amy Gutman. Princeton, N.J.: Princeton University Press, 2001.

Lefort, Claude. *Democracy and Political Theory.* Translated by David Macey. Minneapolis: University of Minnesota Press, 1988.

Lyotard, Jean-Francois. *Toward the Post-Modern.* Atlantic Highlands, N.J.: Humanities Press, 1993.

McGowan, John. *Hannah Arendt: An Introduction.* Minneapolis: University of Minnesota Press, 1998.

———. "Must Politics Be Violent? Arendt's Utopian Vision." In *Hannah Arendt and the Meaning of Politics.* Edited by Craig Calhoun and John McGowan. Minneapolis: University of Minnesota Press, 1997.

Montesquieu. *The Spirit of the Laws.* Translated by Thomas Nugent. London: Haftner Library of Classics, 1949.

Moruzzi, Norma Claire. *Speaking through the Mask: Hannah Arendt and the Politics of Social Identity.* Ithaca, N.Y.: Cornell University Press, 2000.

Nancy, Jean-Luc. *The Birth to Presence.* Stanford, Calif.: Stanford University Press, 1993.

Passerin D'Entreves, Maurizio. *The Political Philosophy of Hannah Arendt.* London: Routledge, 1994.

Patterson, Orlando. *Slavery and Social Death: A Comparative Study.* Cambridge, Mass.: Harvard University Press, 1982.

Rawls, John. *The Law of Peoples.* Cambridge, Mass.: Harvard University Press, 1999.

Rousseau, Jean Jacques. *Discourse on the Origin of Inequality.* Edited by Victor Gourevitch. Cambridge: Cambridge University Press, 1997.

———. *Emile.* Translated by Allen Bloom. New York: Basic Books, 1979.

———. *The Social Contract.* Edited by Victor Gourevitch. Cambridge: Cambridge University Press, 2003.

Sen, Amartya. *Resources, Values, and Development.* Cambridge, Mass.: Harvard University Press, 1984.

Shue, Henry. *Basic Rights: Subsistence, Affluence, and U.S. Foreign Policy.* Princeton, N.J.: Princeton University Press, 1996.

Strauss, Leo. *Natural Right and History.* Chicago, Ill.: University of Chicago Press, 1953.

Taminiaux, Jacques. *The Thracian Maid and the Professional Thinker.* Albany: State University of New York Press, 1997.

Taylor, Charles. *Sources of the Self: The Making of Modern Identity.* Cambridge, Mass.: Harvard University Press, 1989.

Villa, Dana. *Politics, Philosophy, Terror.* Princeton, N.J.: Princeton University Press, 1999.

INDEX

PEG BIRMINGHAM is Professor of Philosophy at DePaul University. She is co-editor (with Philippe van Haute) of *Dissensus Communis: Between Ethics and Politics* and the author of many articles on Arendt, Heidegger, Foucault, and Kristeva that have appeared in such publications as *Research in Phenomenology, Hypatia,* and *The Graduate Journal of Philosophy.* She is co-translator (with Elizabeth Birmingham) of Dominique Janicaud's *Powers of the Rational* (Indiana University Press, 1994).

www.ingramcontent.com/pod-product-compliance
Ingram Content Group UK Ltd.
Pitfield, Milton Keynes, MK11 3LW, UK
UKHW030806140325
456233UK00007B/124